Dimensions Math®
Teacher's Guide PKA

Authors and Reviewers

Tricia Salerno

Pearly Yuen

Jenny Kempe

Dr. Leslie Arceneaux

Allison Coates

Cassandra Turner

Bill Jackson

Singapore Math Inc.

Published by Singapore Math Inc.

19535 SW 129th Avenue
Tualatin, OR 97062
www.singaporemath.com

Dimensions Math Teacher's Guide Pre-Kindergarten A
ISBN 978-1-947226-28-9

First published 2018

Printed in China

Acknowledgments

Copy editing by the Singapore Math Inc. team.
Layout by Cameron Wray with Kelly Barten.
Design and illustration by Cameron Wray.

Contents

Chapter		Lesson	Page

Chapter 7
Numbers to 10 — Part 2

Resources

Dimensions Math® Curriculum

The **Dimensions Math®** series is a Pre-Kindergarten to Grade 5 series based on the pedagogy and methodology of math education in Singapore. The main goal of the **Dimensions Math®** series is to help students develop competence and confidence in mathematics.

The series follows the principles outlined in the Singapore Mathematics Framework below.

Pedagogical Approach and Methodology

- Through "Concrete-Pictorial-Abstract" development, students view the same concepts over time with increasing levels of abstraction.
- Thoughtful sequencing creates a sense of continuity. The content of each grade level builds on that of preceding grade levels. Similarly, lessons build on previous lessons within each grade.
- Group discussion of solution methods encourages expansive thinking.
- Interesting problems and activities provide varied opportunities to explore and apply skills.
- Hands-on tasks and sharing establish a culture of collaboration.
- Extra practice and extension activities encourage students to persevere through challenging problems.
- Variation in pictorial representation (number bonds, bar models, etc.) and concrete representation (straws, linking cubes, base ten blocks, discs, etc.) broadens student understanding.

Each topic is introduced, then thoughtfully developed through the use of a variety of learning experiences, problem solving, student discourse, and opportunities for master of skills. This combination of hands-on practice, in-depth exploration of topics, and mathematical variability in teaching methodology allows students to truly master mathematical concepts.

Singapore Mathematics Framework

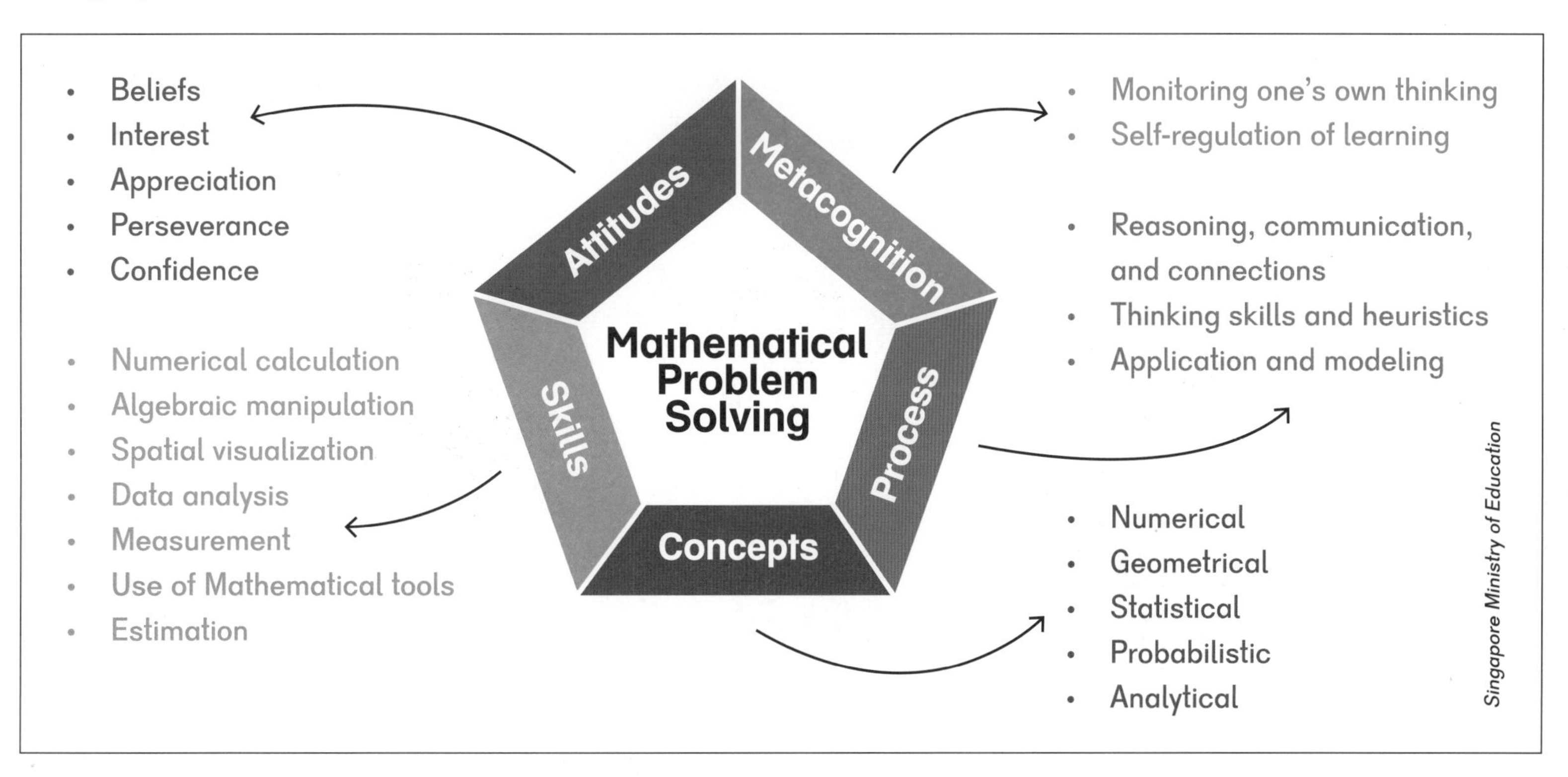

Dimensions Math® Program Materials

Textbooks

Textbooks are designed to help students build a solid foundation in mathematical thinking and efficient problem solving. Careful sequencing of topics, well-chosen problems, and simple graphics foster deep conceptual understanding and confidence. Mental math, problem solving, and correct computation are given balanced attention in all grades. As skills are mastered, students move to increasingly sophisticated concepts within and across grade levels.

Students work through the textbook lessons with the help of five friends: Emma, Alex, Sofia, Dion, and Mei. The characters appear throughout the series and help students develop metacognitive reasoning through questions, hints, and ideas.

A pencil icon at the end of the textbook lessons links to exercises in the workbooks.

Workbooks

Workbooks provide additional problems that range from basic to challenging. These allow students to independently review and practice the skills they have learned.

Teacher's Guides

Teacher's Guides include lesson plans, mathematical background, games, helpful suggestions, and comprehensive resources for daily lessons.

Tests

Tests contain differentiated assessments to systematically evaluate student progress.

Online Resources

The following can be downloaded from dimensionsmath.com.

- **Blackline Masters** used for various hands-on tasks.
- **Letters Home** to be emailed or sent home with students for continued exploration. These outline what the student is learning in math class and offer suggestions for related activities at home. Reinforcement at home supports deep understanding of mathematical concepts.
- **Videos** of popular children songs used for singing activities.
- **Imagery for Projection** from the Chapter Opener and Think activities, for classroom discussion prior to opening the textbooks.
- **Material Lists** for each chapter and lesson, so teachers and classroom helpers can prepare ahead of time.
- **Activities** that can be done with students who need more practice or a greater challenge, organized by concept, chapter, and lesson.
- **Standards Alignments** for various states.

Using the Teacher's Guide

This guide is designed to assist in planning daily lessons, and should be considered a helping hand between the curriculum and the classroom. It provides introductory notes on mathematical content, key points, and suggestions for activities. It also includes ideas for differentiation within each lesson, and answers and solutions to textbook and workbook problems.

Each chapter of the guide begins with the following sections.

- ## Overview

  Includes objectives, vocabulary, and suggested number of class periods for each chapter.

- ## Notes

  Highlights key learning points, explains the purpose of certain activities, and helps teachers understand the flow of topics throughout the year.

- ## Materials

  Lists materials, manipulatives, and Blackline Masters used in the Explore and Learn sections of the guide. It also includes suggested storybooks and snacks. Blackline Masters can be found at dimensionsmath.com.

The guide goes through the Chapter Openers, Daily Lessons, and Practices of each chapter in the following general format.

- ## Explore

  Introduces students to math concepts through hands-on activities. Depending on the classroom, students may be in a circle for Explore then transition to their tables for Learn.

- ## Learn

  Summarizes the main concepts of the lesson, including exercises from Look and Talk pages of the corresponding textbook pages.

Whole Group Play & Small Group Center Play

Allows students to practice concepts through hands-on tasks and games, including suggestions for outdoor play (most of which can be modified for a gymnasium or classroom).

Activities

These recurring activities are for groups of 2 to 4 students at one center. They can be used in math class or at other times of the day.

Sort	**Blocks**	**Art**	**Dramatic Play**
Match	**Counting**	**Patterns**	**Music**

Extend

This expands on **Explore**, **Learn**, and **Play** and provides opportunities for students to deepen their understanding and build confidence.

Discussion is a critical component of each lesson. Have students share their ideas with a partner, small group, or the class as often as possible. As each classroom is different, this guide does not anticipate all situations. Teachers are encouraged to elicit higher level thinking and discussion through questions like these:

- Why? How do you know?
- Can you explain that?
- Can you draw a picture of that?
- Does your answer make sense? How do you know?
- How is this task like the one we did before? How is it different?
- What did you learn before that can help you to solve this problem?
- What is alike and what is different about that?
- Can you solve that a different way?
- How do you know it's true?
- Can you restate or say in your own words what your classmate shared?

Lesson structures and activities do not have to conform exactly to what is shown in the guide. Teachers are encouraged to exercise their discretion in using this material in a way that best suits their classes.

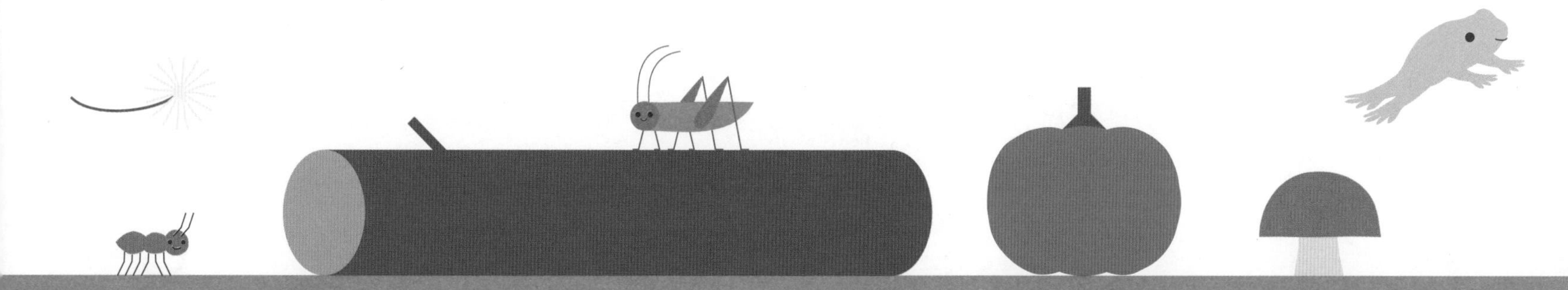

Dimensions Math® Scope & Sequence

PKA

Chapter 1
Match, Sort, and Classify

Red and Blue
Yellow and Green
Color Review
Soft and Hard
Rough, Bumpy, and Smooth
Sticky and Grainy
Size — Part 1
Size — Part 2
Sort Into Two Groups
Practice

Chapter 2
Compare Objects

Big and Small
Long and Short
Tall and Short
Heavy and Light
Practice

Chapter 3
Patterns

Movement Patterns
Sound Patterns
Create Patterns
Practice

Chapter 4
Numbers to 5 — Part 1

Count 1 to 5 — Part 1
Count 1 to 5 — Part 2
Count Back
Count On and Back
Count 1 Object
Count 2 Objects
Count Up to 3 Objects
Count Up to 4 Objects
Count Up to 5 Objects
How Many? — Part 1
How Many? — Part 2
How Many Now? — Part 1
How Many Now? — Part 2
Practice

Chapter 5
Numbers to 5 — Part 2

1, 2, 3
1, 2, 3, 4, 5 — Part 1
1, 2, 3, 4, 5 — Part 2
How Many? — Part 1
How Many? — Part 2
How Many Do You See?
How Many Do You See Now?
Practice

Chapter 6
Numbers to 10 — Part 1

0
Count to 10 — Part 1
Count to 10 — Part 2
Count Back
Order Numbers
Count Up to 6 Objects
Count Up to 7 Objects
Count Up to 8 Objects
Count Up to 9 Objects
Count Up to 10 Objects — Part 1
Count Up to 10 Objects — Part 2
How Many?
Practice

Chapter 7
Numbers to 10 — Part 2

6
7
8
9
10
0 to 10
Count and Match — Part 1
Count and Match — Part 2
Practice

PKB

Chapter 8
Ordinal Numbers

First
Second and Third
Fourth and Fifth
Practice

Chapter 9
Shapes and Solids

Cubes, Cylinders, and Spheres
Cubes
Positions
Build with Solids
Rectangles and Circles
Squares
Triangles

Dimensions Math® Scope & Sequence

Counting Up to 10 Things — Part 2
Recognize the Numbers 6 to 10
Write the Numbers 6 and 7
Write the Numbers 8, 9, and 10
Write the Numbers 6 to 10
Count and Write the Numbers 1 to 10
Ordinal Positions
One More Than
Practice

Chapter 4
Shapes and Solids

Curved or Flat
Solid Shapes
Closed Shapes
Rectangles
Squares
Circles and Triangles
Where is It?
Hexagons
Sizes and Shapes
Combine Shapes
Graphs
Practice

Chapter 5
Compare Height, Length, Weight, and Capacity

Comparing Height
Comparing Length
Height and Length — Part 1
Height and Length — Part 2
Weight — Part 1
Weight — Part 2
Weight — Part 3
Capacity — Part 1
Capacity — Part 2
Practice

Chapter 6
Comparing Numbers Within 10

Same and More
More and Fewer
More and Less
Practice — Part 1
Practice — Part 2

KB

Chapter 7
Numbers to 20

Ten and Some More
Count Ten and Some More
Two Ways to Count
Numbers 16 to 20
Number Words 0 to 10
Number Words 11 to 15
Number Words 16 to 20
Number Order
1 More Than or Less Than
Practice — Part 1
Practice — Part 2

Chapter 8
Number Bonds

Putting Numbers Together — Part 1
Putting Numbers Together — Part 2
Parts Making a Whole
Look for a Part
Number Bonds for 2, 3, and 4
Number Bonds for 5
Number Bonds for 6
Number Bonds for 7
Number Bonds for 8
Number Bonds for 9
Number Bonds for 10
Practice — Part 1
Practice — Part 2
Practice — Part 3

Chapter 9
Addition

Introduction to Addition — Part 1
Introduction to Addition — Part 2
Introduction to Addition — Part 3
Addition
Count On — Part 1
Count On — Part 2
Add Up to 3 and 4
Add Up to 5 and 6
Add Up to 7 and 8
Add Up to 9 and 10
Addition Practice
Practice

Chapter 10
Subtraction

Take Away to Subtract — Part 1

Dimensions Math® Scope & Sequence

Compare Numbers to 20
Addition
Subtraction
Practice

Chapter 6
Addition to 20

Add by Making 10 — Part 1
Add by Making 10 — Part 2
Add by Making 10 — Part 3
Addition Facts to 20
Practice

Chapter 7
Subtraction Within 20

Subtract from 10 — Part 1
Subtract from 10 — Part 2
Subtract the Ones First
Word Problems
Subtraction Facts Within 20
Practice

Chapter 8
Shapes

Solid and Flat Shapes
Grouping Shapes
Making Shapes
Practice

Chapter 9
Ordinal Numbers

Naming Positions
Word Problems
Practice
Review 2

1B

Chapter 10
Length

Comparing Lengths Directly
Comparing Lengths Indirectly
Comparing Lengths with Units
Practice

Chapter 11
Comparing

Subtraction as Comparison
Making Comparison
Subtraction Stories
Picture Graphs
Practice

Chapter 12
Numbers to 40

Numbers to 40
Tens and Ones
Counting by Tens and Ones
Comparing
Practice

Chapter 13
Addition and Subtraction Within 40

Add Ones
Subtract Ones
Make the Next Ten
Use Addition Facts
Subtract from Tens
Use Subtraction Facts
Add Three Numbers
Practice

Chapter 14
Grouping and Sharing

Adding Equal Groups
Sharing
Grouping
Practice

Chapter 15
Fractions

Halves
Fourths
Practice
Review 3

Chapter 16
Numbers to 100

Numbers to 100
Tens and Ones
Count by Ones or Tens
Compare Numbers to 100
Practice

Chapter 17
Addition and Subtraction Within 100

Add Ones — Part 1
Add Tens
Add Ones — Part 2
Add Tens and Ones — Part 1
Add Tens and Ones — Part 2
Subtract Ones — Part 1
Subtract from Tens
Subtract Ones — Part 2
Subtract Tens

Dimensions Math® Scope & Sequence

Dividing by 2
Dividing by 5 and 10
Practice C
Word Problems
Review 2

2B

Chapter 8
Mental Calculation

Adding Ones Mentally
Adding Tens Mentally
Making 100
Adding 97, 98, or 99
Practice A
Subtracting Ones Mentally
Subtracting Tens Mentally
Subtracting 97, 98, or 99
Practice B
Practice C

Chapter 9
Multiplication and Division of 3 and 4

The Multiplication Table of 3
Multiplication Facts of 3
Dividing by 3
Practice A
The Multiplication Table of 4
Multiplication Facts of 4
Dividing by 4
Practice B
Practice C

Chapter 10
Money

Making $1
Dollars and Cents
Making Change
Comparing Money
Practice A
Adding Money
Subtracting Money
Practice B

Chapter 11
Fractions

Halves and Fourths
Writing Unit Fractions
Writing Fractions
Fractions that Make 1 Whole
Comparing and Ordering Fractions
Practice
Review 3

Chapter 12
Time

Telling Time
Time Intervals
A.M. and P.M.
Practice

Chapter 13
Capacity

Comparing Capacity
Units of Capacity
Practice

Chapter 14
Graphs

Picture Graphs
Bar Graphs
Practice

Chapter 15
Shapes

Straight and Curved Sides
Polygons
Semicircles and Quarter-circles
Patterns
Solid Shapes
Practice
Review 4
Review 5

3A

Chapter 1
Numbers to 10,000

Numbers to 10,000
Place Value — Part 1
Place Value — Part 2
Comparing Numbers
The Number Line
Practice A
Number Patterns
Rounding to the Nearest Thousand
Rounding to the Nearest Hundred
Rounding to the Nearest Ten
Practice B

Dimensions Math® Scope & Sequence

Practice B
Review 2

3B

Chapter 9
Fractions — Part 1

Fractions of a Whole
Fractions on a Number Line
Comparing Fractions with Like Denominators
Comparing Fractions with Like Numerators
Practice

Chapter 10
Fractions — Part 2

Equivalent Fractions
Finding Equivalent Fractions
Simplifying Fractions
Comparing Fractions — Part 1
Comparing Fractions — Part 2
Practice A
Adding and Subtracting Fractions — Part 1
Adding and Subtracting Fractions — Part 2
Practice B

Chapter 11
Length

Meters and Centimeters
Subtracting from Meters
Adding Lengths in Meters and Centimeters
Subtracting Lengths in Meters and Centimeters
Practice A
Kilometers
Subtracting from Kilometers
Adding Lengths in Kilometers and Meters
Subtracting Kilometers and Meters
Practice B

Chapter 12
Geometry

Circles
Angles
Right Angles
Triangles
Properties of Triangles
Properties of Quadrilaterals
Practice

Chapter 13
Area and Perimeter

Area
Units of Area
Area of Rectangles
Area of Composite Figures
Practice A
Perimeter
Perimeter of Rectangles
Area and Perimeter
Practice B
Review 3

Chapter 14
Time

Units of Time
Calculating Time — Part 1
Practice A
Calculating Time — Part 2
Calculating Time — Part 3
Calculating Time — Part 4
Practice B

Chapter 15
Money

Dollars and Cents
Making $10
Adding Money
Subtracting Money
Word Problems
Practice

Chapter 16
Capacity and Weight

Liters and Milliliters
Adding and Subtracting Liters and Milliliters
Kilograms and Grams
Adding and Subtracting Kilograms and Grams
Practice

Chapter 17
Graphs and Tables

Picture Graphs and Bar Graphs
Bar Graphs and Tables
Practice
Review 4
Review 5

Suggested number of class periods: 10 – 11

Lesson		Page	Resources	Objectives
	Chapter Opener	p. 5	TB: p. 1	
1	Red and Blue	p. 6	TB: p. 2 WB: p. 1	Recognize the colors red and blue.
2	Yellow and Green	p. 9	TB: p. 5 WB: p. 3	Recognize the colors yellow and green. Classify by appearance and color.
3	Color Review	p. 11	TB: p. 7 WB: p. 5	Sort and classify objects by color. Review colors red, blue, yellow, and green.
4	Soft and Hard	p. 13	TB: p. 9 WB: p. 7	Sort and classify objects by texture. Learn the words “soft” and “hard.”
5	Rough, Bumpy, and Smooth	p. 15	TB: p. 11 WB: p. 9	Classify objects by texture. Learn the words “rough,” “bumpy,” and “smooth.”
6	Sticky and Grainy	p. 17	TB: p. 13 WB: p. 11	Classify objects by texture. Learn the words “sticky” and “grainy.”
7	Size — Part 1	p. 19	TB: p. 15 WB: p. 13	Recognize the color orange. Learn the words “big,” “little,” “small,” and “large.” Match objects by size.
8	Size — Part 2	p. 21	TB: p. 17 WB: p. 15	Classify objects by size and color. Review size. Practice fine motor skills.
9	Sort Into Two Groups	p. 23	TB: p. 19 WB: p. 17	Sort objects into two groups and justify the sort.
10	Practice	p. 25	TB: p. 21 WB: p. 19	Practice concepts introduced in this chapter.
	Workbook Solutions	p. 26		

Chapter Vocabulary

- Same
- Alike
- Not the same
- Different
- Red
- Blue
- Yellow
- Green
- Orange
- Sort
- Match
- Soft
- Hard
- Smooth
- Rough
- Bumpy
- Grainy
- Sticky
- Big
- Little
- Small
- Large
- Size

Notes

Chapter 1 introduces students to important ideas in mathematics. The ability to match, sort, and classify is critical to a child's understanding of the world. Observing similarities and differences helps children categorize. As children develop, their ability to detect slight differences continues to develop through matching, sorting, and categorizing.

Pre-K students have been conducting informal matching, sorting, and classifying since infancy. They learned that when they smiled, the person looking at them would often smile back.

Key Points

Students may have different ideas about how to match and sort. Allow them to explain their reasoning. Always encourage creative thinking.

Encourage use of math vocabulary. If a Pre-K student can identify and name a Tyrannosaurus rex, he or she can use developmentally appropriate vocabulary.

You may want to invite student to bring small rocks to class for lesson activities.

Materials

- Classroom objects – some identical to others, some similar, and some unique
- Tub of objects that can be sorted
- Objects – different shades of red and blue
- Crayons – red and blue
- Objects – different shades of yellow and green
- Yellow objects – 6 of more, two of which are identical
- Crayons – yellow and green
- Objects – red, blue, yellow, green, and some other colors
- Containers – one of each color per center, including red, blue, yellow, green, and transparent
- Objects – hard and soft, at least 1 per student
- Small bags to hold materials, 1 bag of each per pair of students
- Burlap
- Sandpaper
- Bumpy objects – e.g. bubble wrap, woven baskets, pine cones, LEGO® bricks
- Smooth objects – e.g. polished rocks, pieces of satin, plastic counting chips
- Pieces of wood to sand
- Containers of rice, salt, sand, and dough
- Glue sticks and tape
- 2 orange objects that look and feel the same but are different sizes
- 3 objects that are the same except for their sizes
- Objects that are the same except for color (red, blue, yellow, green, and orange only) and size, such as:
  - a large red ball and a small yellow ball
  - a large blue crayon and a small orange crayon
  - a large green sock and a small red sock
- Bins, 1 per 4 students
- Sets of objects – to be sorted by size, color, or feel
- Bins, 1 per 4 students

Note: Materials for Activities will be listed in detail in each lesson.

Blackline Masters

- Rhyming Rebus
- Who Needs Teeth?
- Table Setting

Storybooks

- *Rocks Hard, Soft, Smooth, and Rough* by Natalie M. Rosinsky
- *Goldilocks and the Three Bears*
- Possible touch and feel books for this chapter:
  - *Pat the Bunny* by Dorothy Kunhardt
  - *Tails* by Matthew Van Fleet
  - *Fuzzy Fuzzy Fuzzy!* by Sandra Boynton

Optional Snacks

- Red berries
- Blueberries
- Bananas
- Celery sticks
- Green peas (fresh or frozen)
- Kiwifruit
- Pretzels (soft twists and hard sticks)
- Dried fruit
- Apple slices
- Crackers, smooth and textured, large and small

Letters Home

- Chapter 1 Letter

Notes

Lesson Materials

- Classroom objects – some identical to others, some similar, and some unique
- Tub of objects that can be sorted

Explore

Place numerous objects on the floor. Some of the objects should be the same, some should be similar, and some should be unique. From these objects, ask students to choose two that are the same and two that are different, as you introduce those words by showing two objects that are the same and two that are different. Have students discuss their findings and reasoning.

Give small groups of students a tub of objects to sort into two groups. After students have sorted the objects, have them explain the reasons for the sorts. Have them put the objects back in the tub and try to sort them a different way, again explaining the reasons for their sorts.

Learn

Have students look at page 1 and ask them which things are the same and which are different. The purpose here is to get students to express their thinking. There may be more than one way of observing the same object. For example, students might notice:

- The backpacks are the same because they hold books.
- The backpacks are different because they don't look the same.
- The carpet squares are the same because they are all squares.
- The carpet squares are different because they are different colors.
- The hula hoops are the same.

Extend Learn

That's a Good Question: Ask the students additional questions about the picture. Suggested questions:

- What is alike and what is different about the backpacks?
- What is alike and what is different about the stuffed toys on the table?

Objective

- Recognize the colors red and blue.

Lesson Materials

- Objects — different shades of red and blue
- Crayons — red and blue
- Optional snack: Red berries and blueberries

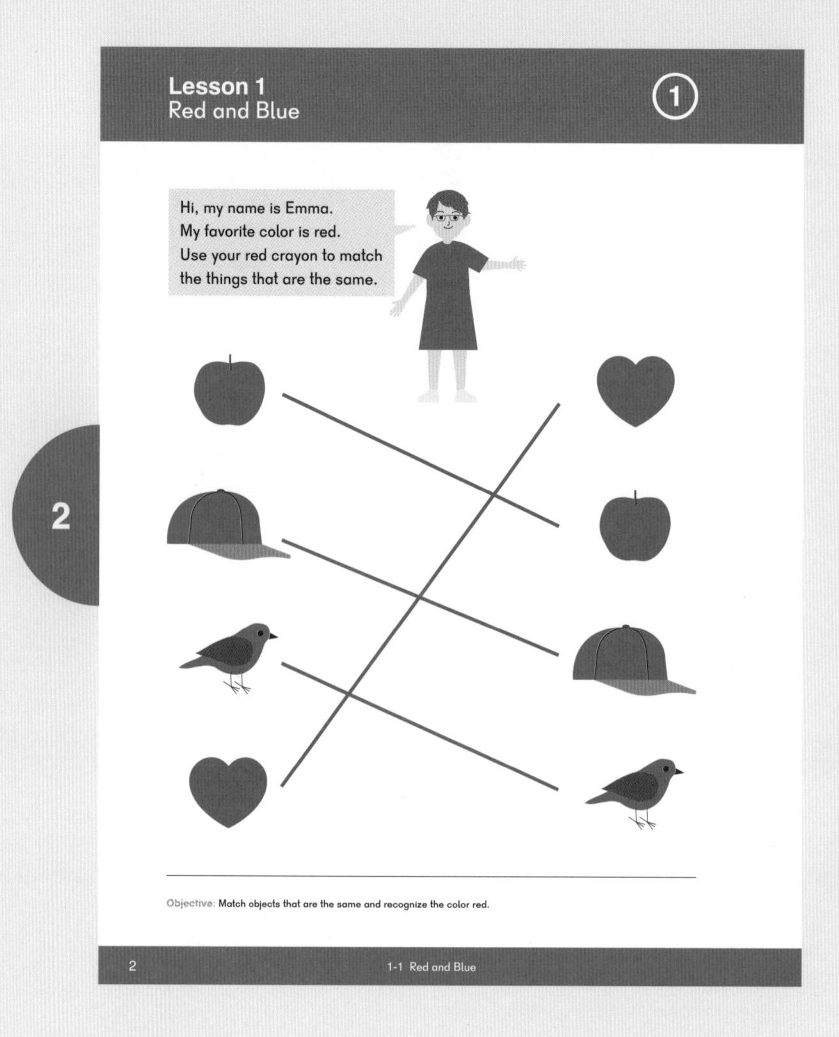

2

Explore

Show students a red object and a blue object (cubes, crayons, or markers work well). Ask if they know the names of the colors.

Discuss the different shades of red and blue. In **Dimensions Math® Pre-K**, all shades of a color will be considered that color. Light blue, royal blue, and navy blue, for example, are all blue. This variation in shade will add to discussion of similarities and differences.

Have several students search the room and bring back objects that are red. Have several other students bring back objects that are blue.

Have students share what they found and say if their object is: (a) the same color and item as the example (red crayon and red crayon); or (b) the same color but a different object (blue crayon and blue book).

Have students return their objects to where they found them. Allow students who haven't had a turn to repeat the activity, except that their task will be to gather objects that are a different color and a different thing from the examples.

Learn

Introduce Emma and discuss her comments. Have students identify which objects in each column are the same. Have students use their fingers to draw a line to match objects at first, then use a red crayon.

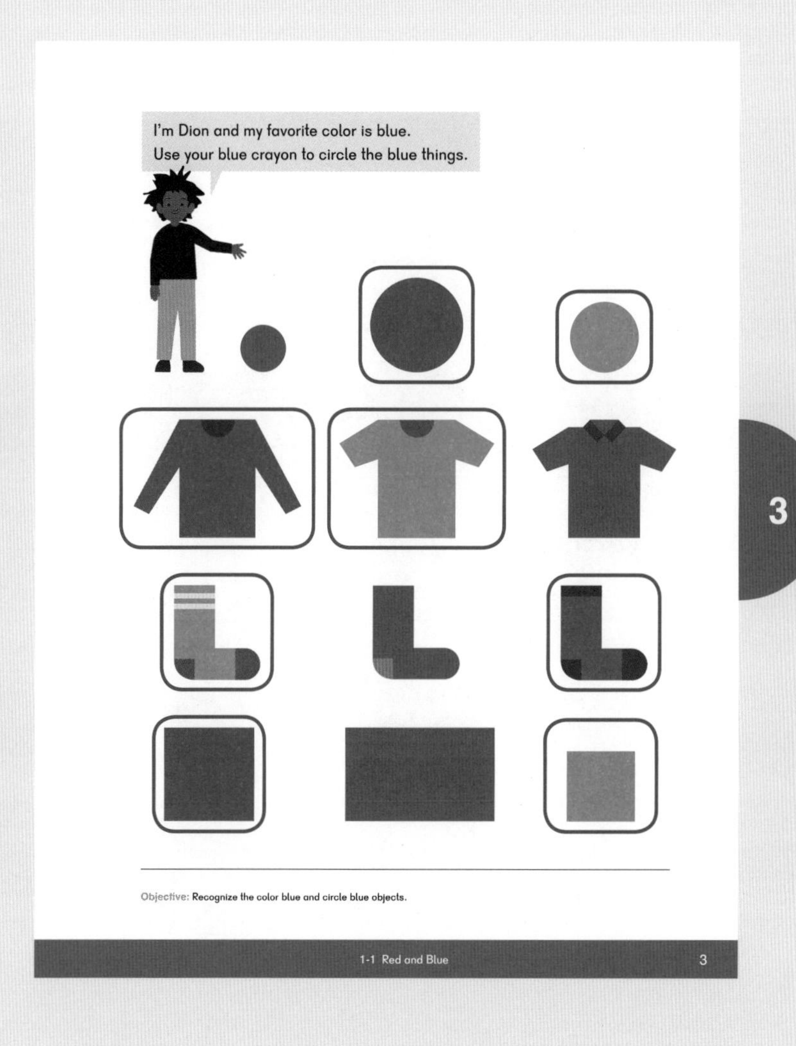

Introduce Dion and discuss his comments. Ask students to name the objects on page 3. What is the same or different about the balls, the shirt, the socks, and the blocks?

Draw a circle in the air with your finger and have students do the same. Draw a circle on the board. Have students use a finger to trace the outside of the circles on the page, then circle the blue objects with a blue crayon.

Read the nursery rhyme to the students. Have them share what they know about sheep and cows. What is the same about sheep and cows? What is different?

Read Dion's comment and ask students why this rhyme is Dion's favorite.

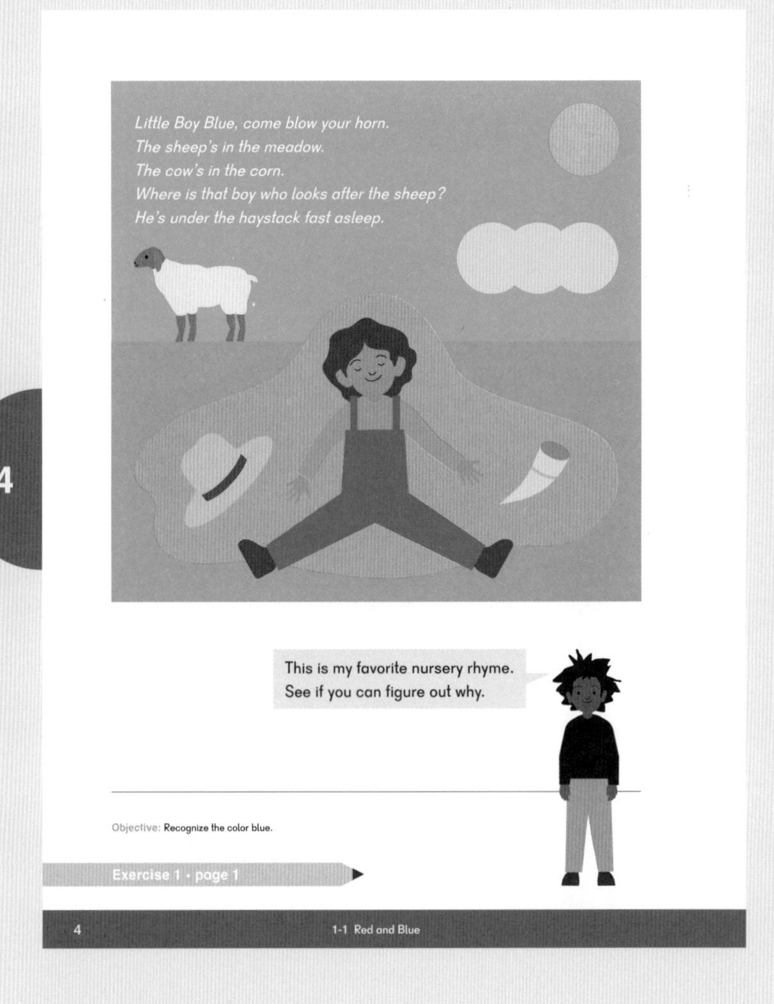

Whole Group Play

I Spy: Play "I Spy" using the colors red and blue as part of the clues.

Stand Up for Your Color: Call, "Red," "Blue," "Not red," or, "Not blue." Students stand if they are wearing the color called.

Small Group Center Play

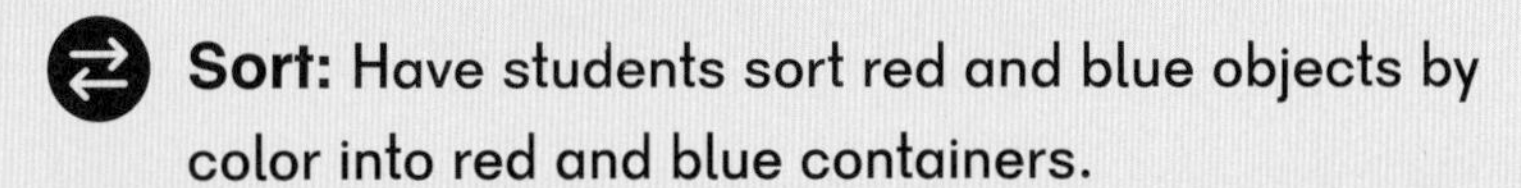

Sort: Have students sort red and blue objects by color into red and blue containers.

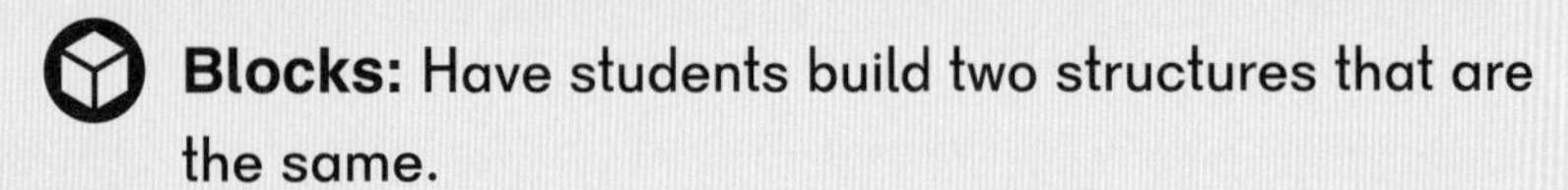

Blocks: Have students build two structures that are the same.

Dress-Up: Include red and blue items of clothing.

Art: Offer red and blue paint for student use.

Exercise 1 • page 1

Extend Learn

Rhyme Time: Give students Rhyming Rebus (BLM) worksheet. Read the words on the page with them and have them make up poems or stories using the rhyming words. Teachers can work with students one-on-one and write down their ideas, or use a recording device.

Materials: Rhyming Rebus (BLM), recording device

Objectives

- Recognize the colors yellow and green.
- Classify objects by appearance and color.

Lesson Materials

- Objects — different shades of yellow and green
- Yellow objects — 6 or more, two of which are identical
- Crayons — yellow and green
- Optional snack: Bananas and celery or frozen peas

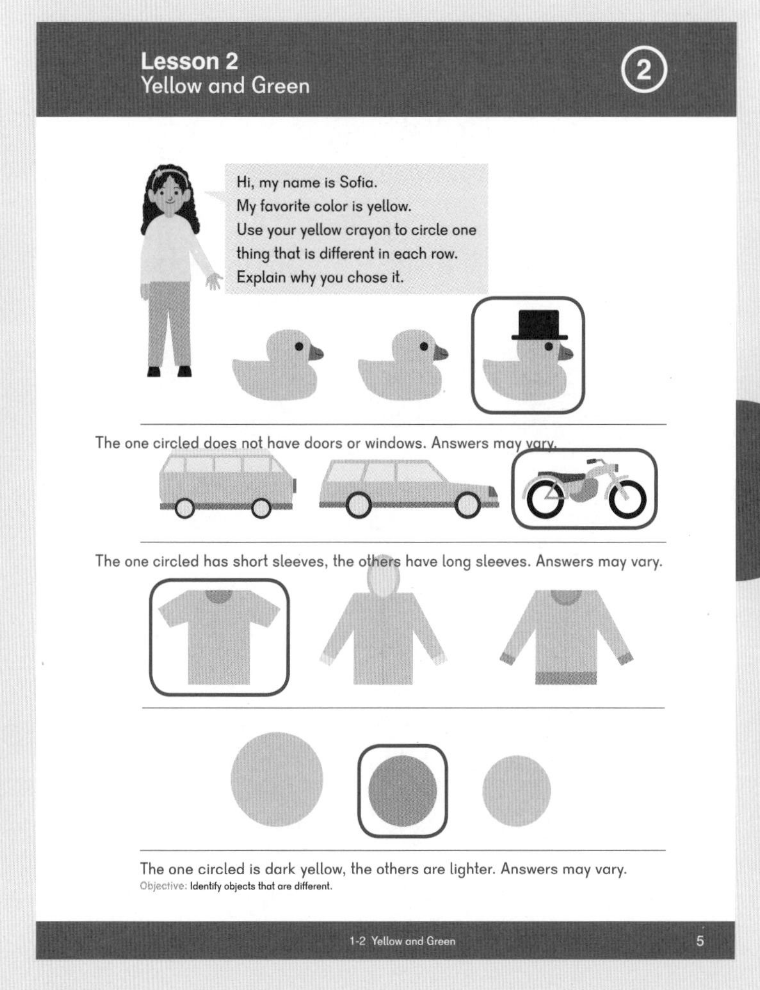

Explore

Ask students what colors they learned about in **Lesson 1: Red and Blue**. Have all students quickly point at something red, then something blue.

Show students a yellow object and a green object. Ask if they know the names of the colors.

Repeat the activity from Lesson 1, but this time students find yellow and green objects. Have students compare the objects they collect to the examples by color and type of object.

Show students three yellow objects, two of which are obviously the same and one of which is different. Ask them which is different and have them say why. To encourage deeper thinking, show another set of three yellow objects and ask the same questions. These three objects, however, may be the same in some ways and different in others so there may be more than one correct answer.

Learn

Introduce Sofia and discuss her comments on page 5. Have students discuss and identify which objects in each row are different and why. Then have them circle the different objects. Note: There may be more than one correct answer. Allow students to explain their thinking.

Introduce Alex and read Sofia's comments.

Ask students to name the objects on page 6 and to say which are yellow and which are green. Discuss same and different again. For example, the two tricycles (or the two yo-yos) are the same except for color.

Whole Group Play

I Spy: Use the colors red, blue, yellow, and green for the clues.

Stand Up for Your Color: Call, "Red," "Blue," "Yellow," "Green," or, "None of the colors." Students stand if they are wearing the color called.

Rubber Ducky: Teach students the song "Rubber Ducky." Teach them how to do the duck walk.

Materials: Rubber Ducky video (VR)

Small Group Center Play

Sort: Have students sort red, blue, yellow, and green objects by color into containers of the correct color.

Blocks: Have students build two structures that are the same.

Dress-Up: Include red, blue, yellow, and green items of clothing.

Art: Offer red, blue, yellow, and green paint for student use.

Exercise 2 • page 3

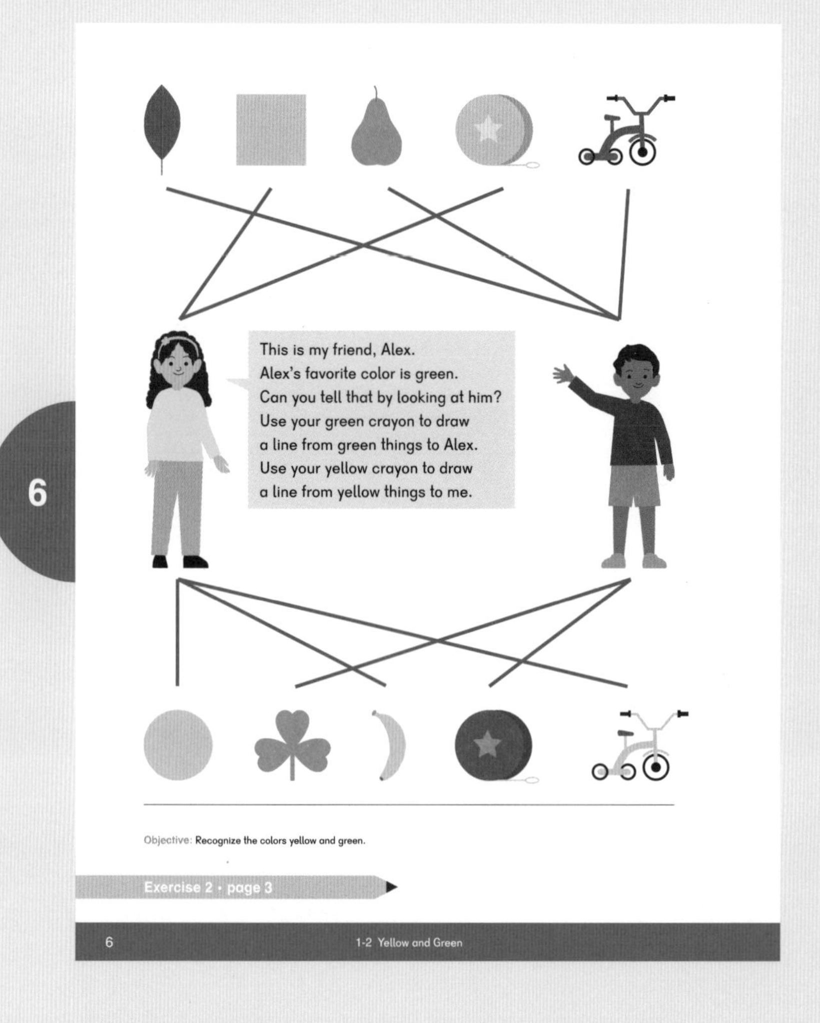

Extend Play

Can You Make What I Made?: Before starting this activity with students, make your own picture using yellow and green crayons and dot stickers. Students will use the same materials to try and recreate your artwork as you describe it, using the words "yellow," "green," "same," and "different."

Materials: Yellow crayons, green crayons, dot stickers, art paper

Objectives

- Sort objects by color.
- Review colors red, blue, yellow, and green.

Lesson Materials

- Objects — red, blue, yellow, green, and some other colors
- Containers — one of each color per center, including red, blue, yellow, green, and transparent
- Optional snack: Red berries, blueberries, bananas, and kiwifruit

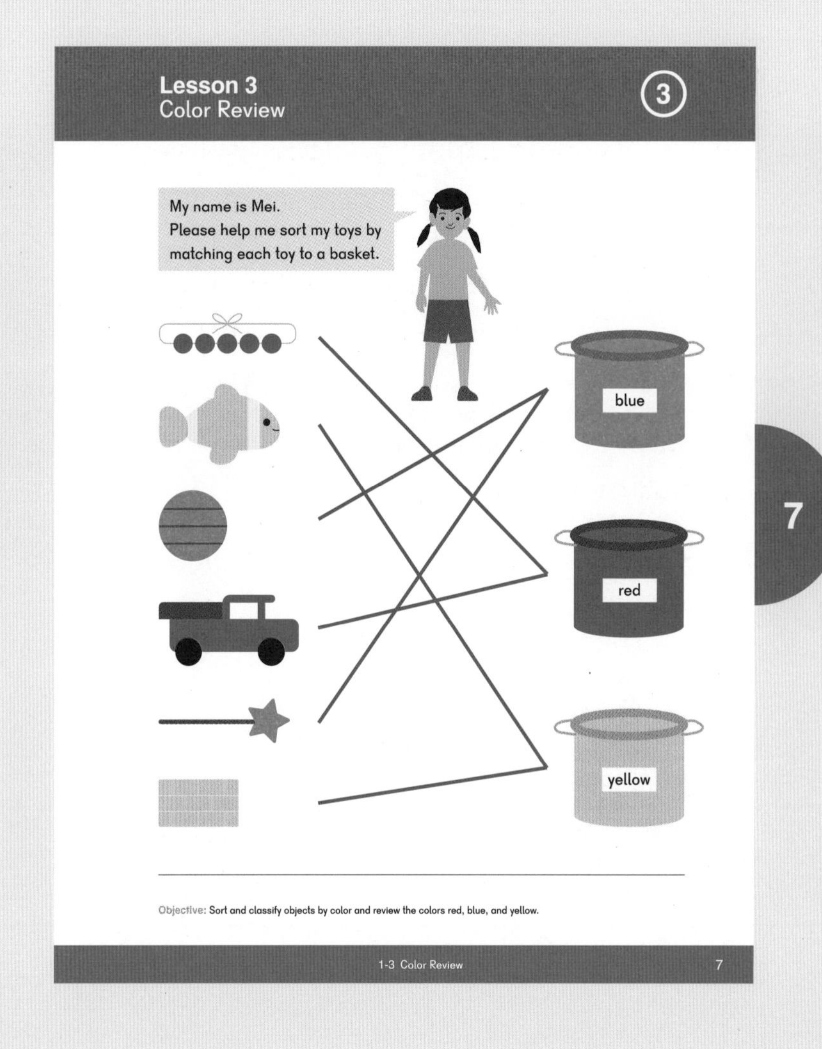

Explore

Ask students to name the colors they have learned in previous lessons. Provide containers and objects that can be sorted into the four colors. Also provide objects that are not one of the learned colors and a transparent container for them. Call on groups of students and tell them to each choose an object and place it in the container they believe it belongs in.

Learn

Review sorted objects and have students agree or disagree with their placement.

Introduce Mei and discuss her comments. Have students complete the matching as an informal assessment.

Whole Group Play

Red Light, Green Light: Hold up a red piece of construction paper or a green piece of construction paper, alternating between the two colors. Identify one student as "It." All students move only on the green and try to touch the student who has been designated "It."

Materials: Red construction paper, green construction paper

I Spy: Play "I Spy" using the four colors introduced as part of the clues.

Small Group Center Play

- **Sort:** Provide red, blue, yellow, and green objects for students to sort by color.
- **Blocks:** Have students build two structures that are the same.
- **Dress-Up:** Include red, blue, yellow, and green items of clothing.
- **Art:** Offer red, blue, yellow, and green paint for student use.

Exercise 3 • page 5

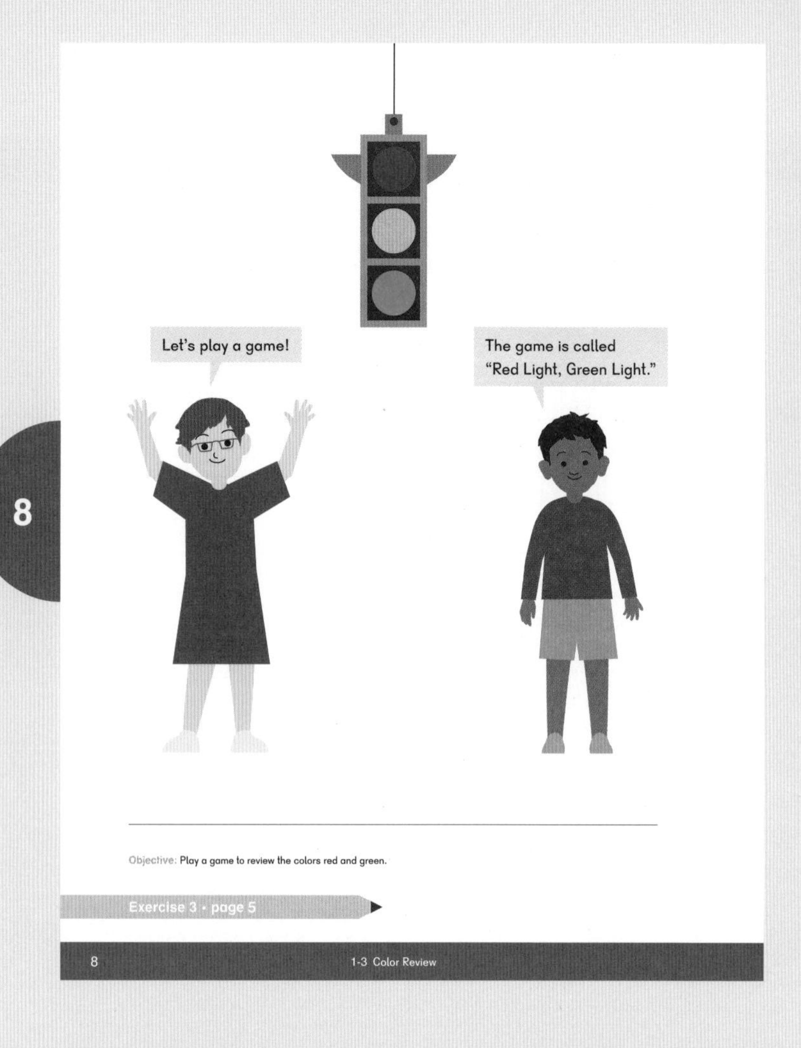

Extend Play

What Was I Thinking?: Have students play in small groups, with one student at a time sorting the other students given their attribute of choice. For example, it could be color of hair, long sleeves vs. short sleeves, etc.

Color Words: Put the words "red," "blue," "yellow," and "green" on your word wall with a picture of something that color.

Objectives

- Sort and classify objects by texture.
- Learn the words "soft" and "hard."

Lesson Materials

- Objects – hard and soft, at least 1 per student
- Optional snack: Soft and hard pretzels

Explore

Provide students with objects that are hard and soft and have them take turns describing them.

Students may choose to describe:

- size
- color
- shape
- uses

Allow for creative thinking.

Ask students to think back to how they described and sorted objects by color. Tell them that today they are going to describe the same objects by how they feel. Guide them to use the words "hard" and "soft."

Learn

Ask students what Dion is holding. Ask them to say which object is hard and which is soft. You may choose to have an example of each for students to feel.

Have students sort their items using hard and soft as their reason. They can put their items into two piles or areas in the classroom.

Read Alex's directions aloud. Have students discuss, then circle the objects they think will be hard.

Whole Group Play

Can You Make What I Made?: Before class, make something using the suggested materials. Have students try to make the same thing.

Materials: Cotton balls, craft sticks, glue

I Spy: Play "I Spy" using the four colors, "hard," and "soft" as part of the clues.

Reading Time: Read books to students and have them take turns describing the pictures they feel using the words "soft" or "hard."

Materials: Touch and feel books (see page 3 of this Teacher's Guide)

Small Group Center Play

Sort: Provide hard and soft objects for students to sort by how the objects feel into separate containers.

Blocks: Add pom-poms to the block center.

Dress-Up: Include red, blue, yellow, green, hard, and soft clothing and accessories.

Art: Include red, blue, yellow, and green paint. Allow students to use sponges and plastic toys to dip into the paint and transfer to art paper.

Exercise 4 • page 7

Extend Play

Chomp, Chomp: Discuss the need for teeth when eating soft and hard foods. Give each student the Who Needs Teeth? (BLM) worksheet and read them directions.

Materials: Who Needs Teeth? (BLM)

Objectives

- Classify objects by texture.
- Learn the words "rough," "bumpy," and "smooth."

Lesson Materials

- Small bags to hold materials, 1 bag of each per pair of students
- Burlap
- Sandpaper
- Bumpy objects — e.g. bubble wrap, woven baskets, pine cones, LEGO® bricks
- Smooth objects — e.g. polished rocks, pieces of satin, plastic counting chips
- Pieces of wood to sand
- Optional snack: Pretzel sticks and apple slices

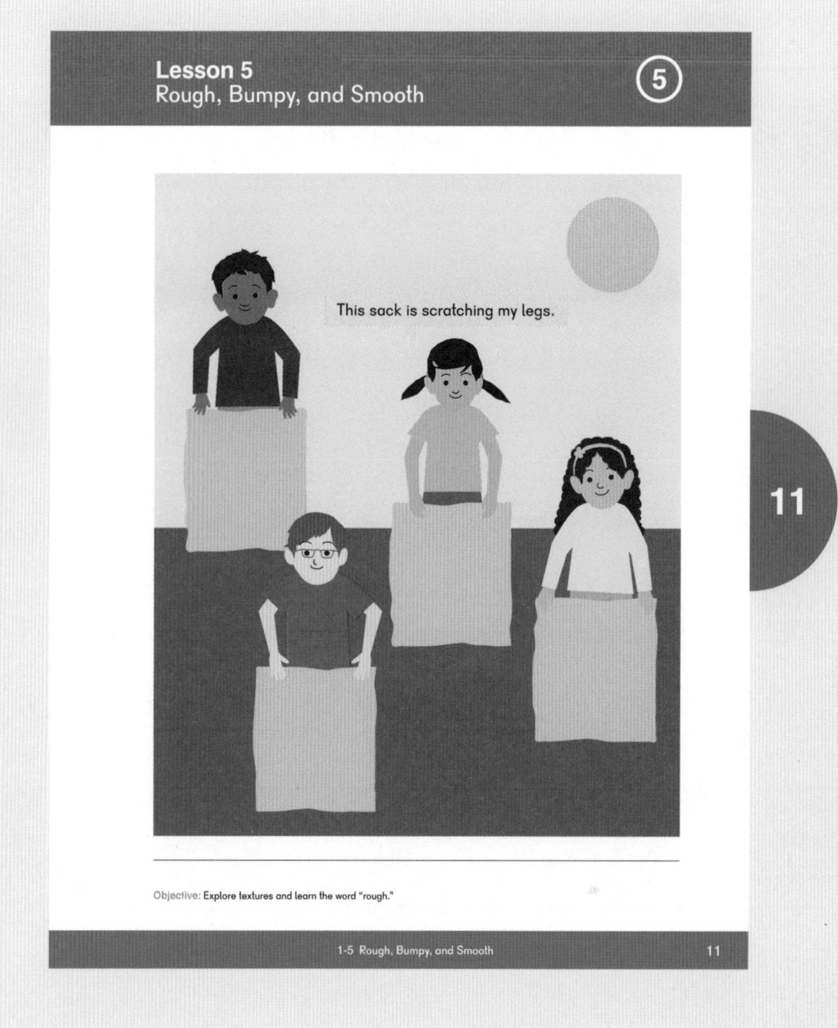

Explore

Fill small bags with rough, bumpy and smooth objects.

Ask students to recall the terms from **Lesson 4: Soft and Hard** which describe how things feel. Ask them if those words describe everything they can feel. Provide pairs of students with their **Explore** bags. Give them time to feel the objects in the bag and discuss with a partner how they might describe the different items. Bring students back to the whole group again and have them share the objects in their bags and describe how they feel. Ask student pairs if their objects feel rough, bumpy, or smooth.

Tell students that just as objects can be described by color, as "hard," or as "soft," they can also be described as rough, bumpy, or smooth. Ask students to name other things in the room that could be rough, bumpy, or smooth. Be aware that some objects may be considered both bumpy and rough. Allow for discussion.

Learn

Ask students if they know what the four friends are doing on page 11. Read Mei's comment. Have them feel a piece of burlap and tell them that the sacks in the illustration are made of that material. Point out that Mei is describing the material in another way when she says, "This sack is scratching my legs." Things that are rough are often scratchy.

Ask students what they think Sofia is doing on page 12. Show how to use sandpaper to sand a piece of wood.

Read Sofia's and Dion's comments aloud. Have students discuss experiences with sandpaper and satin.

Whole Group Play

Sack Race: Give each student both types of fabric sacks. Point out the different textures and show them how to stand in the sacks and hold the top as they jump. Designate a start and finish line and have students race.

Materials: 1 burlap bag and 1 pillowcase per student

Small Group Center Play

Sort: Rough, bumpy, smooth, hard, and soft objects into separate containers.

Blocks: Build with LEGO® or DUPLO® style interlocking blocks.

Dress-Up: Include red, blue, yellow, green, hard, and soft clothing and accessories.

Art: Offer red, blue, yellow, and green paints.

Texture Rub Art: Provide pieces of fine sandpaper, blocks, and bubble wrap. Allow students to decorate their rubbings with stickers, other colors, etc.

Sand It: Have student use a piece of fine sandpaper to sand a piece of wood.

Exercise 5 • page 9

Extend Explore

Reading Time: Have students bring interesting rocks to class. Read *Rocks Hard, Soft, Smooth, and Rough* to students. Have students sort the rocks into groups and give the reasons for their sorts.

Materials: *Rocks Hard, Soft, Smooth, and Rough* by Natalie M. Rosinsky

Objectives

- Classify objects by texture.
- Learn the words "sticky" and "grainy."

Lesson Materials

- Containers of rice, salt, sand, and dough
- Glue sticks and tape
- Optional snack: Treats baked in the lesson or a variety of smooth and textured crackers

Prior to this lesson, set up stations around the room including containers holding rice, salt, and dry sand. Set up other stations with glue sticks, dough, and tape. Be sure that students' hands are dry when they play at the rice, salt, and sand stations.

Explore

Ask students what words they have learned to describe how objects feel.

Tell them that the objects for today's lesson don't fit into hard, soft, rough, bumpy, or smooth categories.

Tell students that they will have a few minutes at each station. When you give the signal, students should move to the next station. Be clear what your signal will be, and in which order the stations are set up. Tell students that while they are at each station, they should think of words to describe how the objects feel.

Note: Students may want to wash and dry their hands between different stations.

Bring students together and have them share descriptions of the objects. Introduce the terms "sticky" and "grainy" and ask them which stations had sticky objects and which had grainy objects.

Learn

Discuss what Alex and Mei are doing on page 13. Read their comments aloud.

Have students identify the objects on page 14. Pointing at the bowl with the spoon, ask them if they have ever made dough. Read Sofia's directions to them and have them complete the task.

Whole Group Play

Little Bakers: Make cookie or bread dough with the students. Bake the dough and enjoy!

Materials: Ingredients for dough, baking supplies

Salt Dough: Mix and knead salt dough.

Materials: Salt, flour, water, mixing bowls, mixing spoons, measuring cups

Small Group Center Play

Sort: Rough, bumpy, smooth, hard, and soft objects into separate containers.

Blocks: Build with LEGO® or DUPLO® style interlocking blocks.

Dress-Up: Include red, blue, yellow, green, hard, and soft clothing and accessories.

Glitter Art: Use glue sticks and glitter to make art.

Sand Art: Make sand art with different colors of sand.

Texture Rub Art: Provide pieces of fine sandpaper, blocks, and bubble wrap. Allow students to decorate their rubbings with stickers, other colors, etc.

Sand It: Have students use a piece of fine sandpaper to sand a piece of wood.

Dry vs. Wet: Wet some sand. Have students compare the feeling of the wet sand to the dry sand. Repeat with rice and salt, if desired.

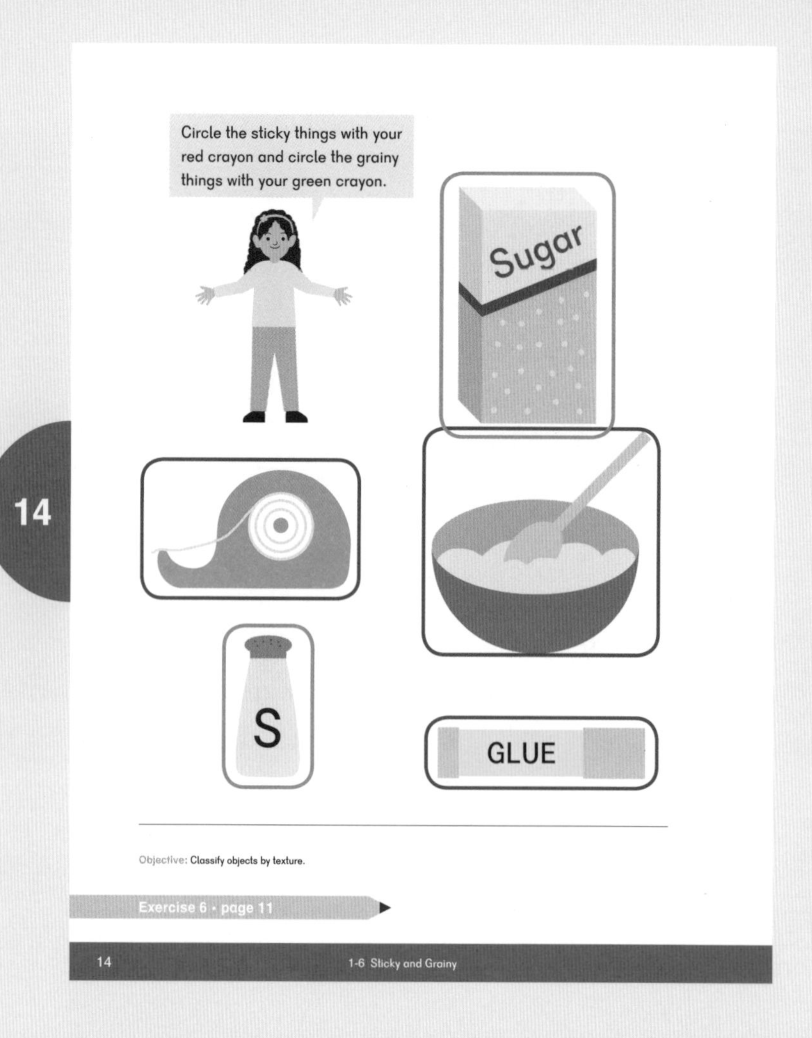

Exercise 6 • page 11

Extend Learn

Rocks and Sand: Set up centers with magnifying glasses, rocks, and sand. Have students explore and describe what they find. You could ask, "How are rocks and sand similar? Where do you think sand comes from?"

Materials: Magnifying glasses, rocks, sand

DUPLO® and LEGO® are trademarks of the LEGO Group of companies.

Objectives

- Recognize the color orange.
- Learn the words "big," "little," "small," and "large."
- Match objects by size.

Lesson Materials

- 2 orange objects that look and feel the same but are different sizes
- 3 objects that are the same except for their sizes
- *Goldilocks and the Three Bears*
- Optional snack: Small and large apple slices

Explore

Ask students how they have learned to describe objects so far. Elicit answers regarding color and how objects feel, including hard, soft, rough, bumpy, smooth, sticky, grainy, and scratchy.

Introduce the color orange.

Show students two orange objects that look and feel the same but are different sizes. Examples: carrots, orange blocks, orange boxes, and/or tangerines.

Ask students how they could describe the objects. Students should notice that the objects are little (or small) and big (or large).

Show students three objects that are the same except they are three different sizes. Ask which is big and which is small. Ask if another term could be used to describe a size "in the middle."

Help students develop visualization skills by having them think about the door in a doll house and a door in the classroom. Which is big? Which is small? Then think about a classroom door and a garage door. Ask them which door is big, small, and in the middle.

Learn

Read *Goldilocks and the Three Bears* to the students, then have them discuss the picture in the textbook. Ask, "What is big, what is small, and what is orange?"

Read Mei's speech bubble to the students and have them draw lines to match objects.

Whole Group Play

Act out Those Sizes: Have students take turns acting out Goldilocks being discovered and running out of the house.

Materials: Big and small chairs for the actors

Mother May I?: Require use of "big" and "small."

Small Group Center Play

Sort: Objects by color and size.

Blocks: Build with LEGO® or DUPLO® style interlocking blocks.

Dress-Up: Include clothes for children and adults.

Kitchen: Provide large and small dishes and bowls for students to use.

Art: Have big and small pieces of paper for students to use. Add orange paint.

Texture Rub Art: At each center, have students rub one big and one small object and have them describe the objects by size, color, and feel. Allow students to decorate their rubbings with stickers, other colors, etc.

Dry vs. Wet: Wet some sand. Have students compare the feeling of the wet sand to the dry sand. Repeat with rice and salt, if desired.

Exercise 7 • page 13

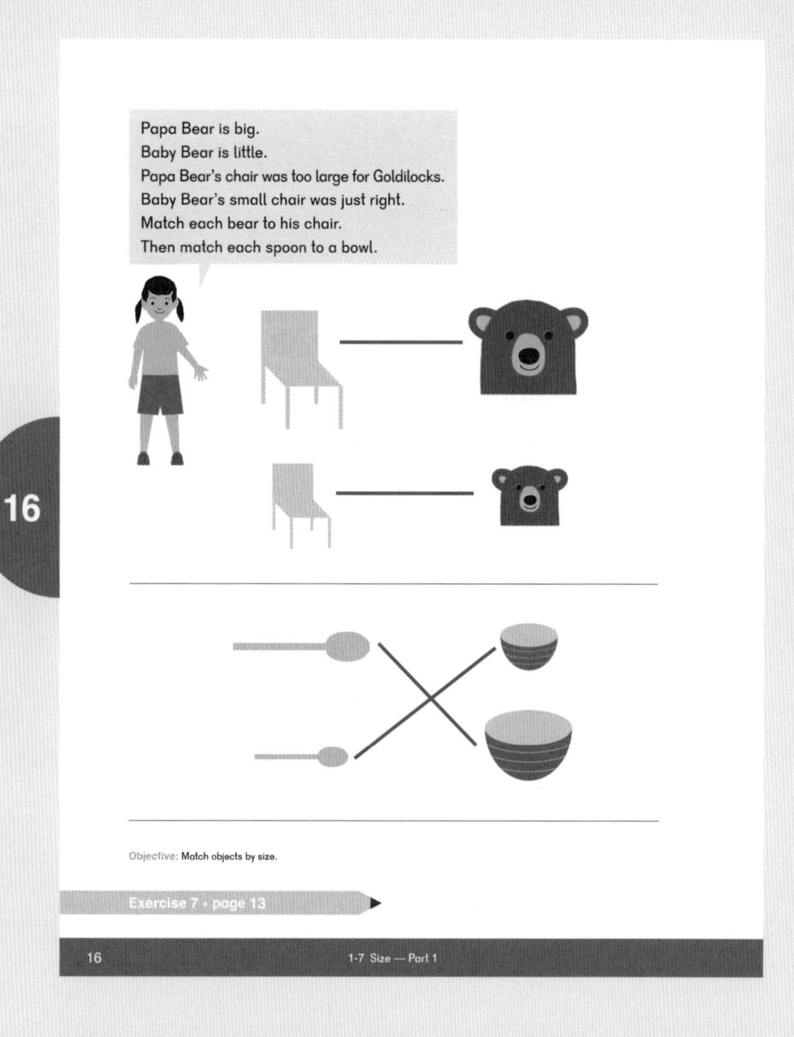

Extend Explore

Is Bigger Better?: Have students discuss times when it is better to have a large object than a similar small object, and times when it is better to have a small object than a similar large object. For example, it is better to have a small spoon to feed a baby, but it is better to have a large spoon to stir food cooking in a pot.

Objectives

- Classify objects by size and color.
- Review size.

Lesson Materials

- Objects that are the same except for color (red, blue, yellow, green, and orange only) and size, such as:
  - a large red ball and a small yellow ball
  - a large blue crayon and a small orange crayon
  - a large green sock and a small red sock
- Bins, 1 per 4 students
- Optional snack: Large and small crackers

17

Prior to this lesson, put together bins of objects that can be sorted by size or color, one bin for each group of four students.

Explore

Discuss the different terms students know to describe items, including size. Tell students that they will work in small groups to sort the items in their bin as they wish. Give students adequate time in their groups to decide on a sort reason and then sort.

Bring students back together and have the groups share how they sorted their objects.

Learn

Have students identify the objects in Emma's bedroom, mentioning size and color. For example, the green picture is big and the red picture is small.

Read Emma's directions to the students. Have them first use their fingers to trace around each given shape. Then have them draw a big shape next to each.

Whole Group Play

Act out Those Sizes: Have students take turns eating the bears' porridge. Require use of "big" and "small" as students describe the objects.

Materials: Bowls and spoons of three different sizes

Mother May I?: Require use of "big" and "small."

Small Group Center Play

Sort: Objects by color and size.

Blocks: Build with LEGO® or DUPLO® style interlocking blocks.

Dress-Up: Include clothes for children and adults.

Kitchen: Provide large and small dishes and bowls for students to use.

Hearts Galore: Give each student a large heart-shaped piece of paper, several smaller heart-shaped pieces of paper, and a glue stick to create a piece of art using the large and small hearts. Encourage them to use as many paper hearts as desired in their creation, and decorate with orange paint, stickers, glitter, markers, crayons, etc.

Texture Rub Art: At each center, have students rub one big and one small object and have them describe the objects by size, color, and feel. Allow students to decorate their rubbings with stickers, other colors, etc.

I Can't See It, but I Can Feel It: Prior to class, cut a hole in the side of the box that is large enough for a student to put their arm through and pull things out of the box. Prepare large/small pairs of items such as an oversized pencil and a short pencil. Place the smaller item from each pair in the box.

18

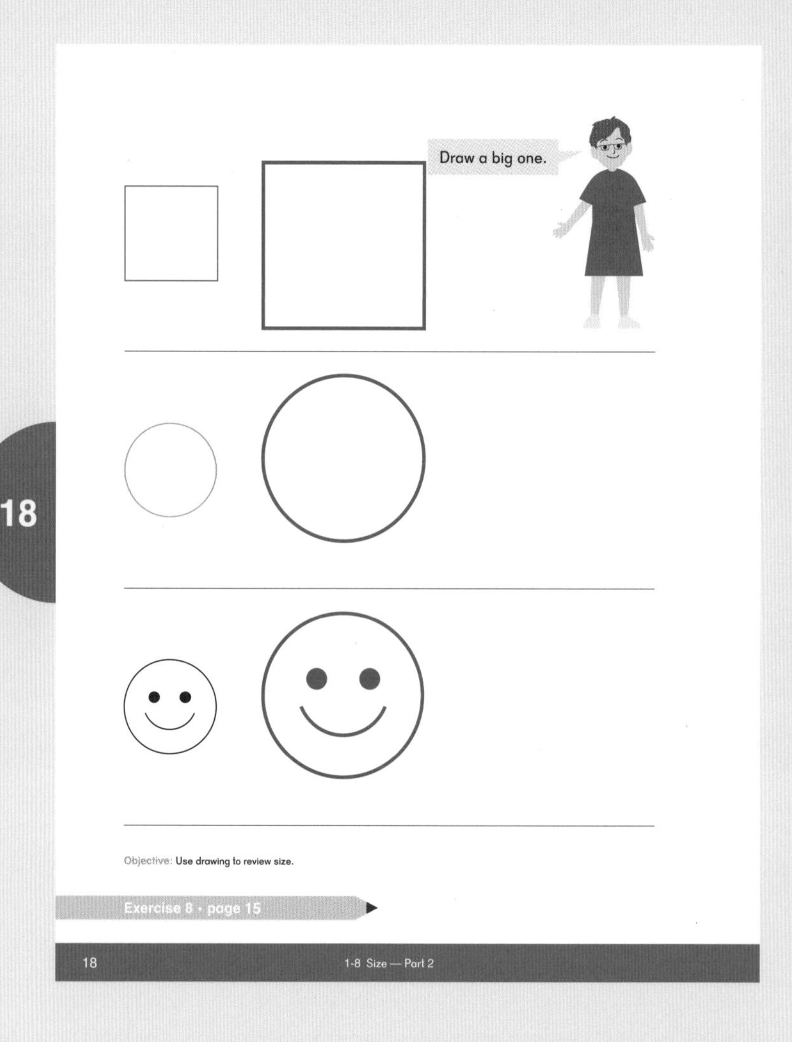

Blindfold a student and hand her a large object such as a ball, block, sock, cup, etc. She will reach into the container and try to find the small object that is similar to the large object she is holding. Repeat until all students have had a turn.

Exercise 8 • page 15

Extend Play

I Can't See It, but I Can Feel It Reverse: Reverse student-teacher roles for the activity **I Can't See It, but I Can Feel It**.

Materials: Cardboard box with hole(s) cut into the side, blindfold, objects in large/small pairs

Lesson 9 Sort Into Two Groups

Objective

- Sort into two groups and justify the sort.

Lesson Materials

- Sets of objects – to be sorted by size, color, or feel
- Bins, 1 per 4 students
- Optional snack: Dried fruit and crackers (given to students in 1 cup with 2 empty cups for them to sort into)

Lesson 9
Sort Into Two Groups
9

I have to sort these things into two groups.

Objective: Discuss reasons for sorting objects into two groups.

1-9 Sort Into Two Groups 19

19

Prior to this lesson, put together bins of objects that can be sorted by size, color, or feel — one bin for each group of four students.

Explore

Tell students that today they will work in small groups to sort the items in their bins as they wish. After they have sorted one way and explained the sort reason, have them mix the objects again and sort them another way. Give students adequate time in their groups to decide on sort reasons and to sort.

Bring students back together and have the groups share how they sorted their objects.

Learn

Have students look at the objects in Dion's toy bucket and discuss how he sorted his toys. Ask students if they see a different way he could have sorted them.

DUPLO® and LEGO® are trademarks of the LEGO Group of companies.

Have students name the objects on page 20. Read Dion's directions to them. Ask them which box is for forks and which is for spoons, encouraging them to use colors to describe the boxes. Have students complete the task.

Whole Group Play

Continue to have students sort and describe objects by playing any of the games from prior lessons.

Small Group Center Play

Sort: Set up centers with different things to sort into two groups such as two types or shapes of buttons, shells, beads, dried beans, etc. Have students work in pairs. One student will sort a set of objects and the other student will guess the reason for the sort. Then reverse their roles.

Blocks: Set up centers with LEGO® bricks or other building blocks in different sizes and colors. First, have students sort the blocks at least one way and say their sorting attribute. Then, allow students to build wonderful creations with their blocks.

Dress-Up: Include clothes for children and adults.

Kitchen: Provide large and small dishes and bowls for students to use.

Set the Table: Give each student a Table Setting (BLM). Have them set the table the same way, using only play materials. If time allows, have students dress up and pretend to eat a meal or have a tea party on their set table.

Hearts Galore: Give each student a large heart-shaped piece of paper, several smaller heart-shaped pieces of paper, and a glue stick to create a piece of art using the large and small hearts. Encourage them to use as many paper hearts as desired in their creation, and decorate with orange paint, stickers, glitter, markers, crayons, etc.

20

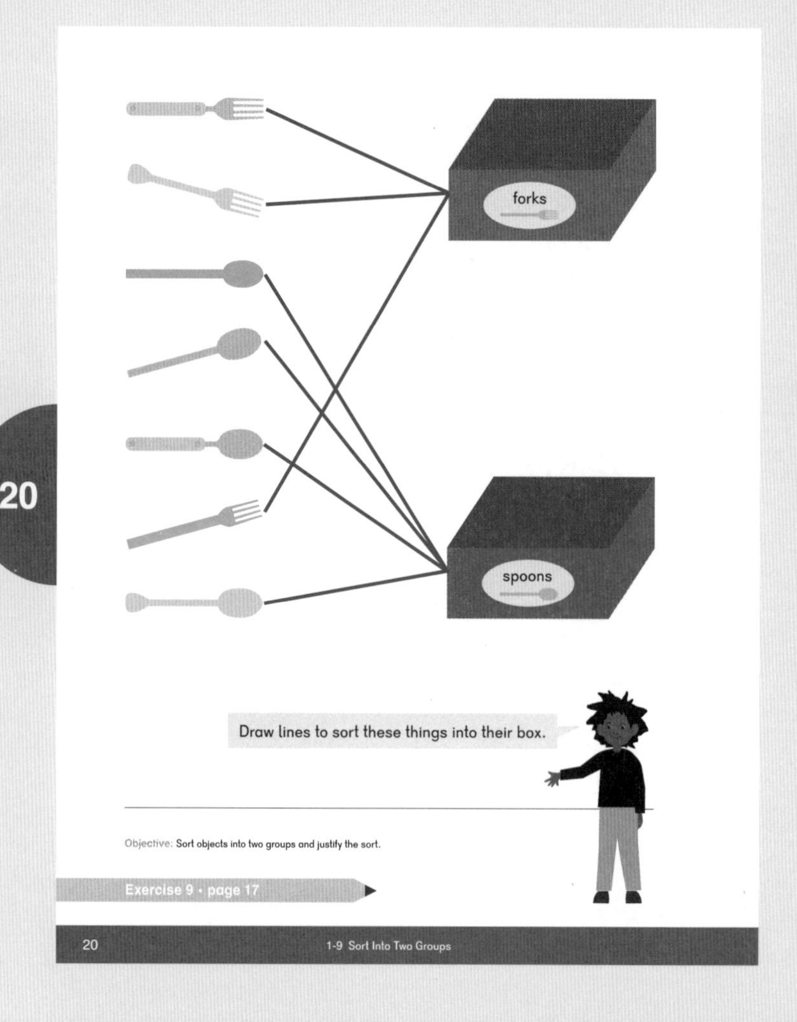

Exercise 9 • page 17

Extend Learn

Kitchen Sort: Have students sort kitchen items in as many ways as possible, explaining the reasoning. Have them try to name other objects which could be added to the sorted groups.

Materials: Kitchen items such as spatulas, bowls, flatware, and towels

Objective

- Practice concepts introduced in this chapter.

Activity Materials

- Optional snack: Variety of fruits and crackers

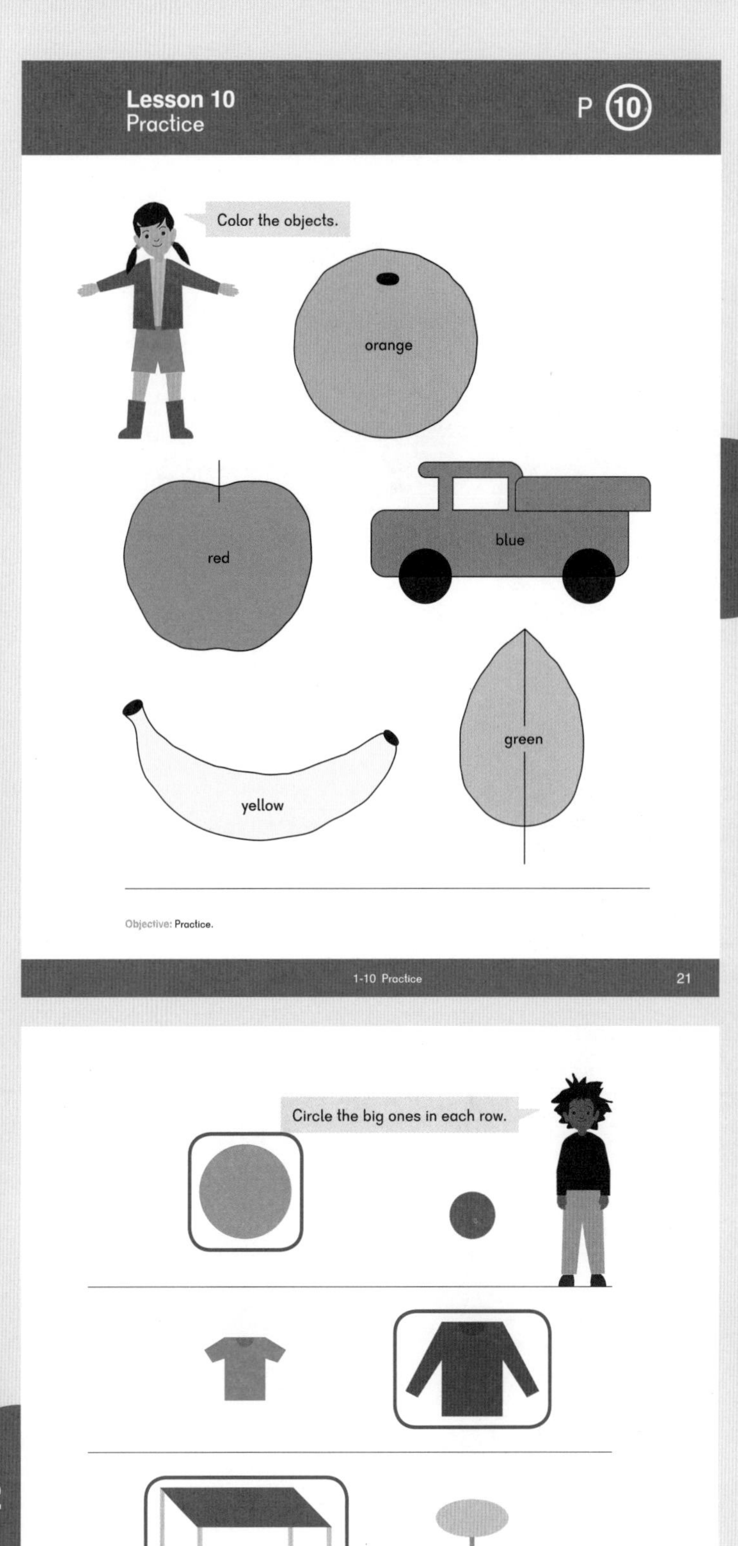

For the **Practice,** read the directions and speech bubbles on each page and have students complete the task.

Exercise 10 • page 19

Extend Explore

What Feels the Same?: Have students try to think of other objects that could be added to a collection of rough, smooth, grainy, or sticky objects.

Exercise 1 • pages 1 – 2

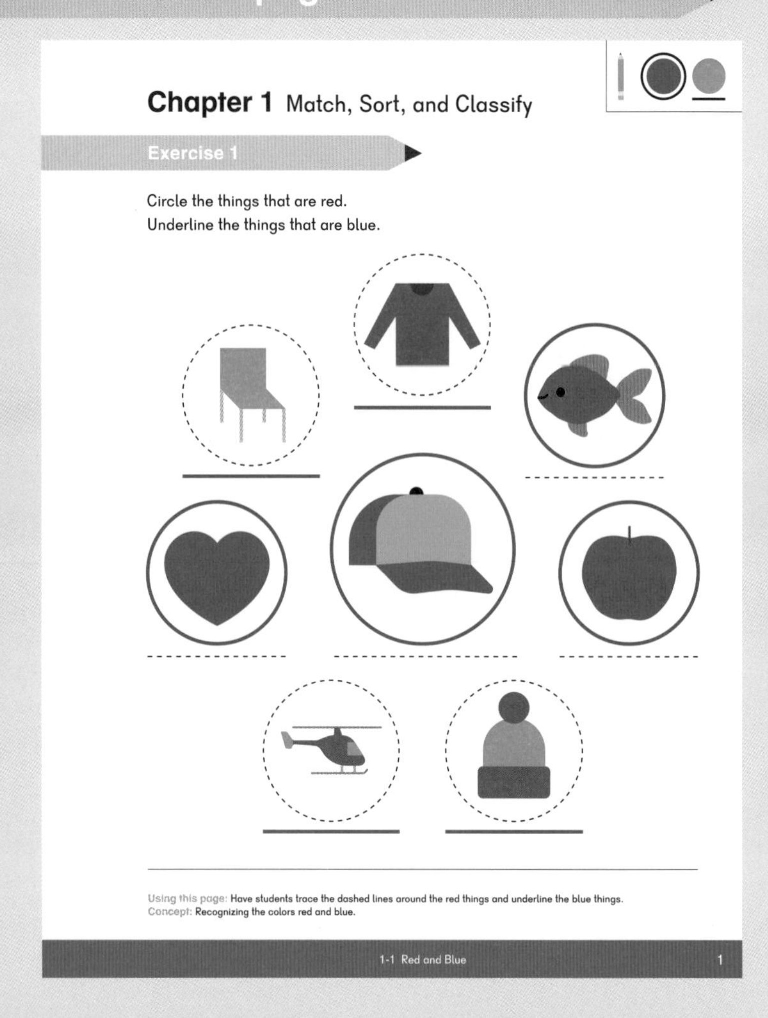

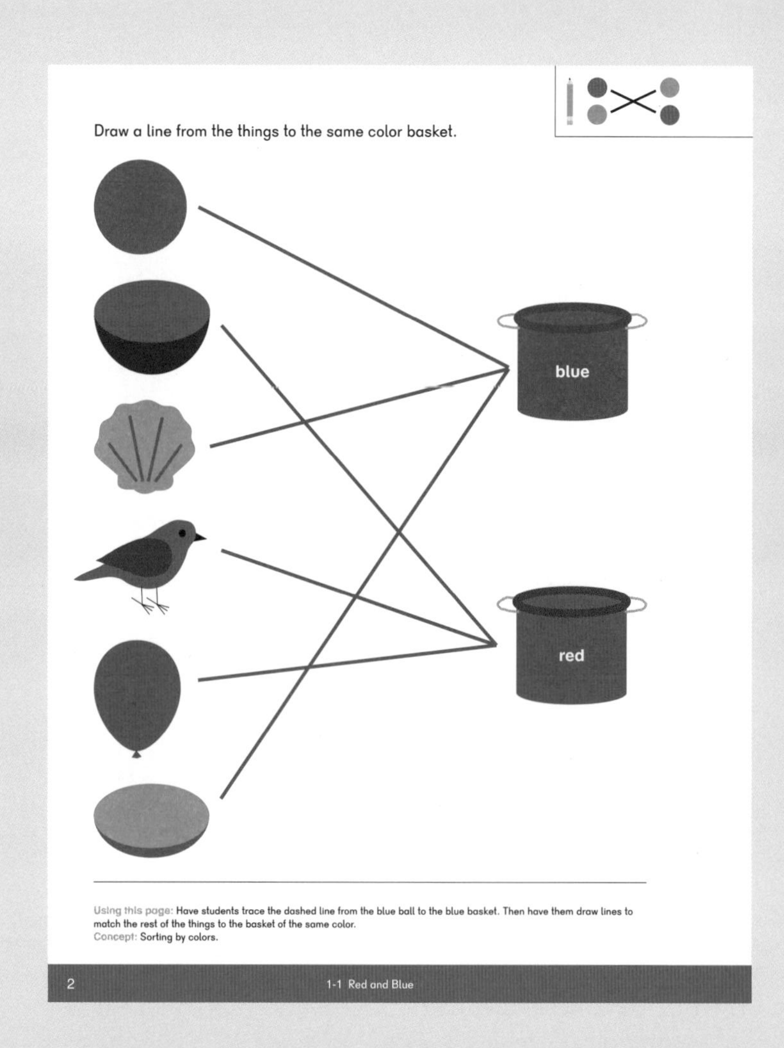

Exercise 2 • pages 3 – 4

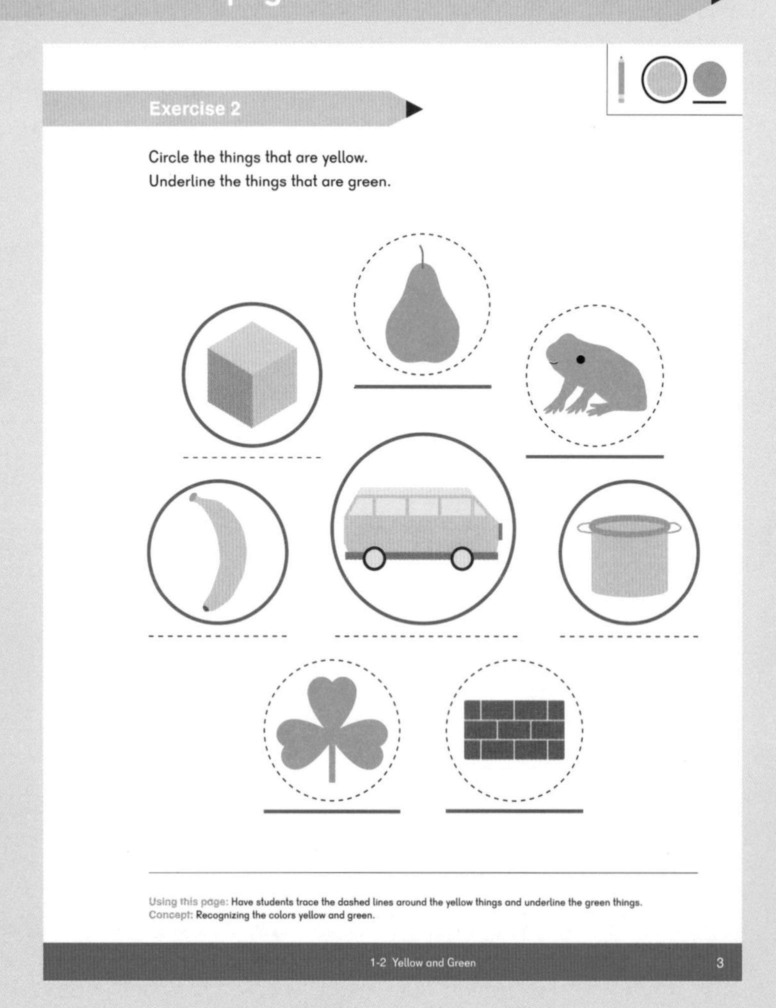

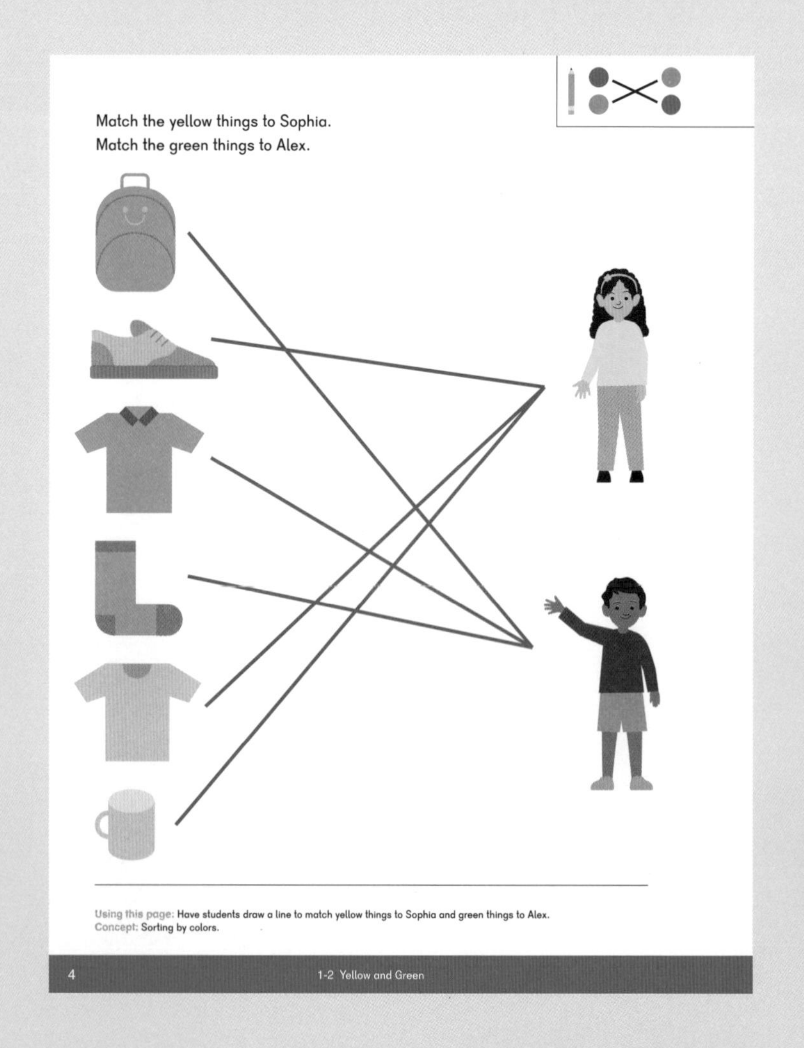

Exercise 3 • pages 5 – 6

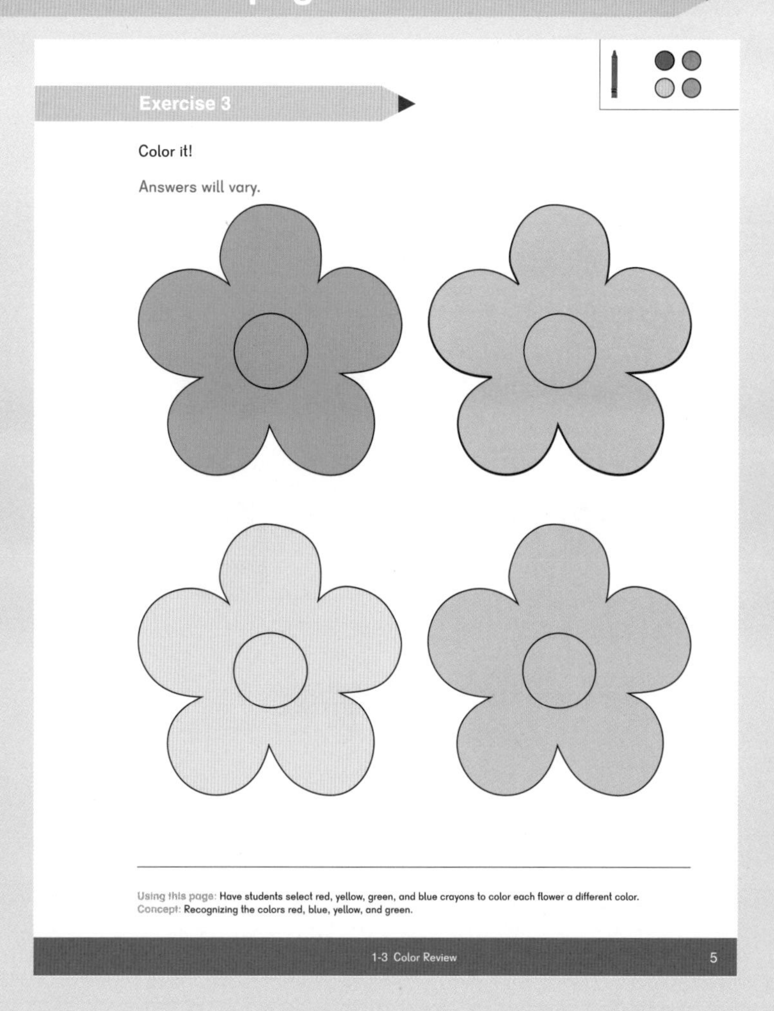

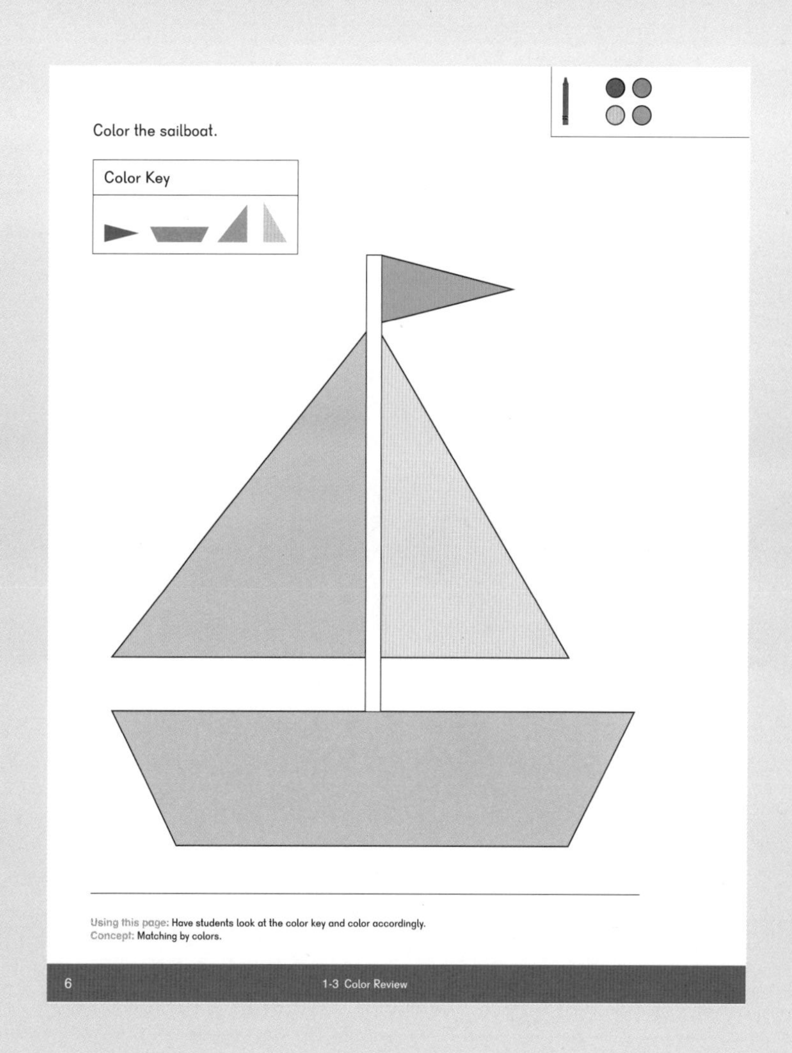

Exercise 4 • pages 7 – 8

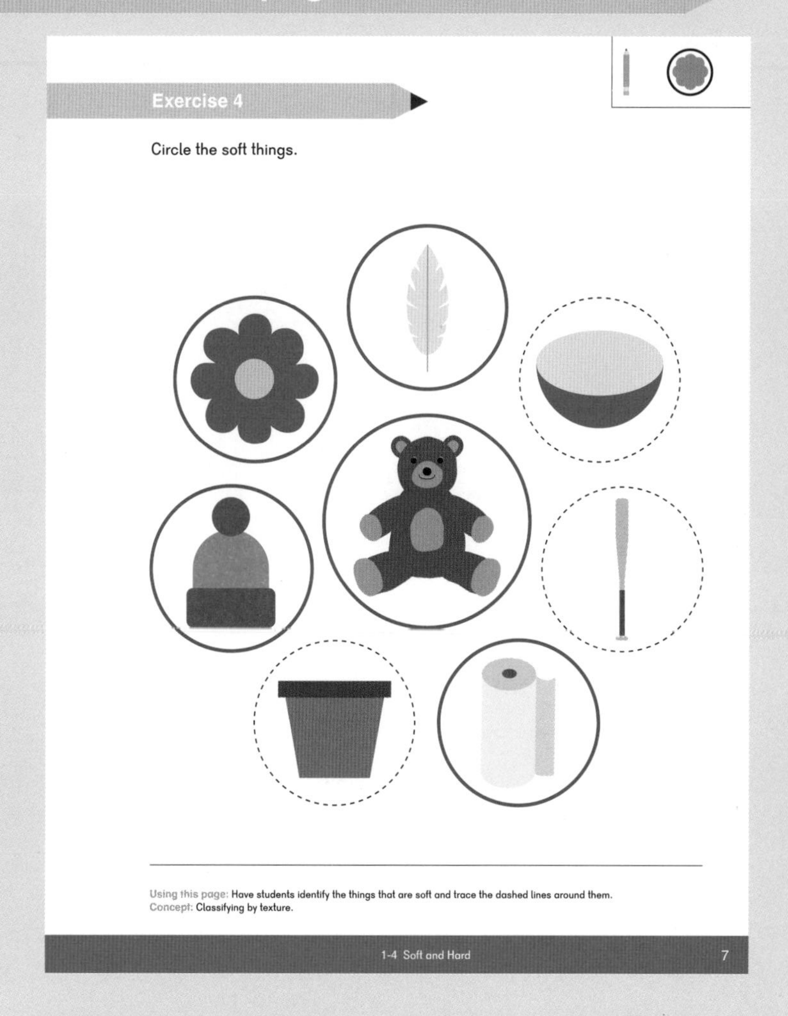

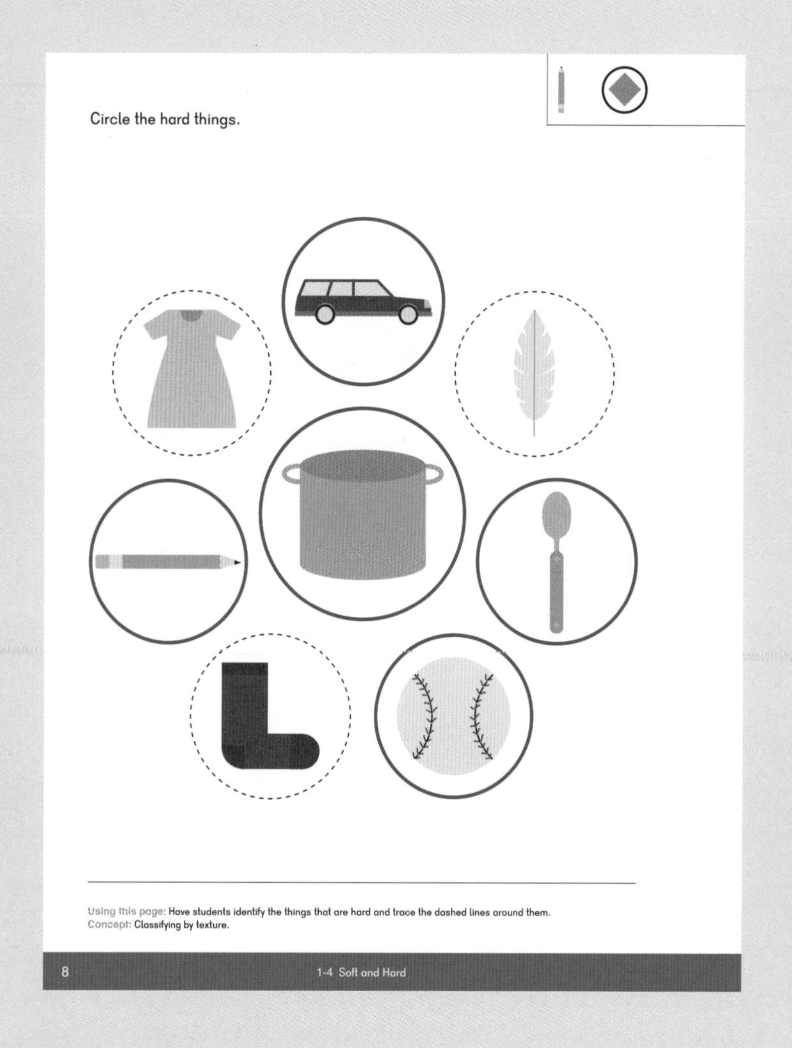

Exercise 5 • pages 9 – 10

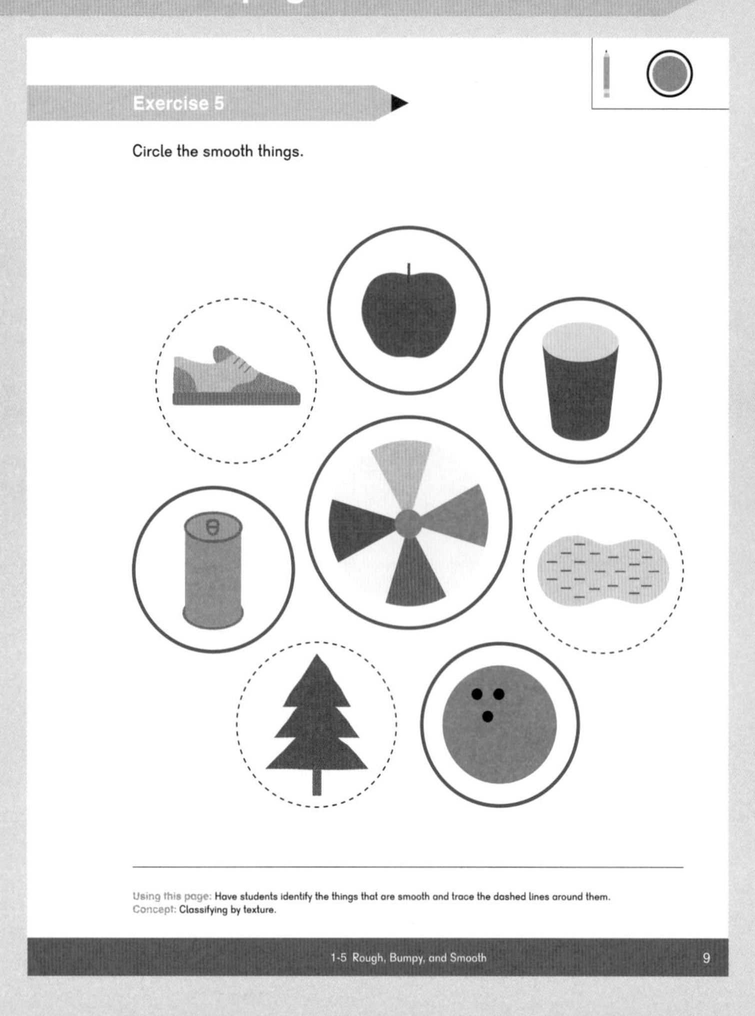

Exercise 6 • pages 11 – 12

Exercise 7 • pages 13 – 14

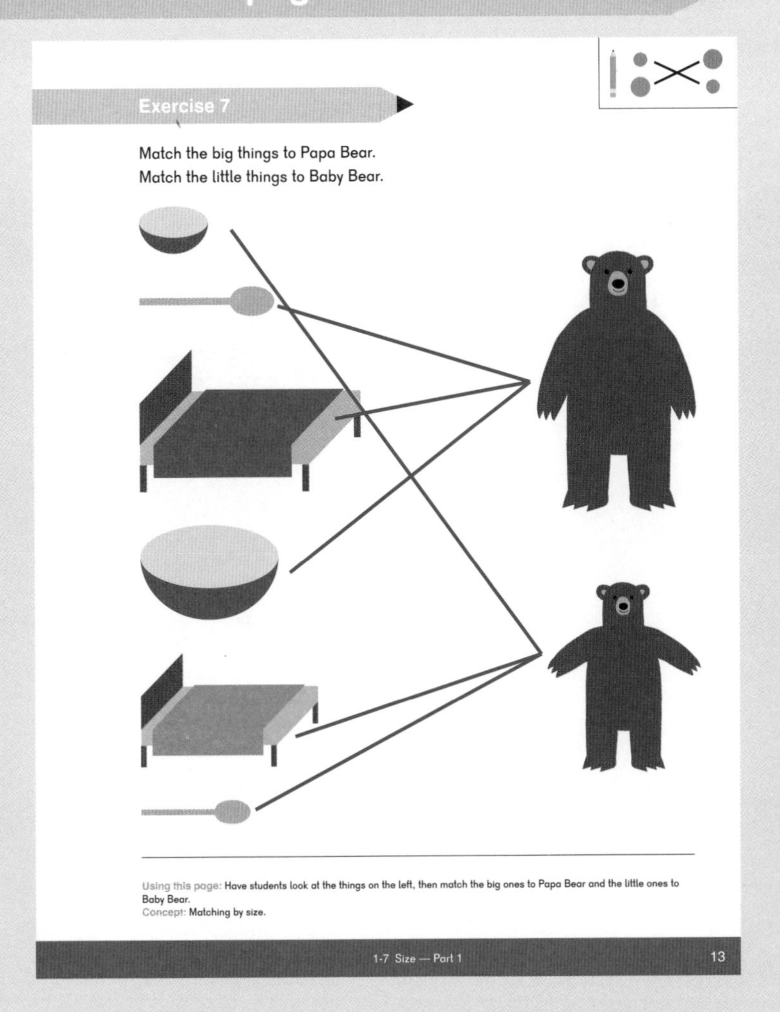

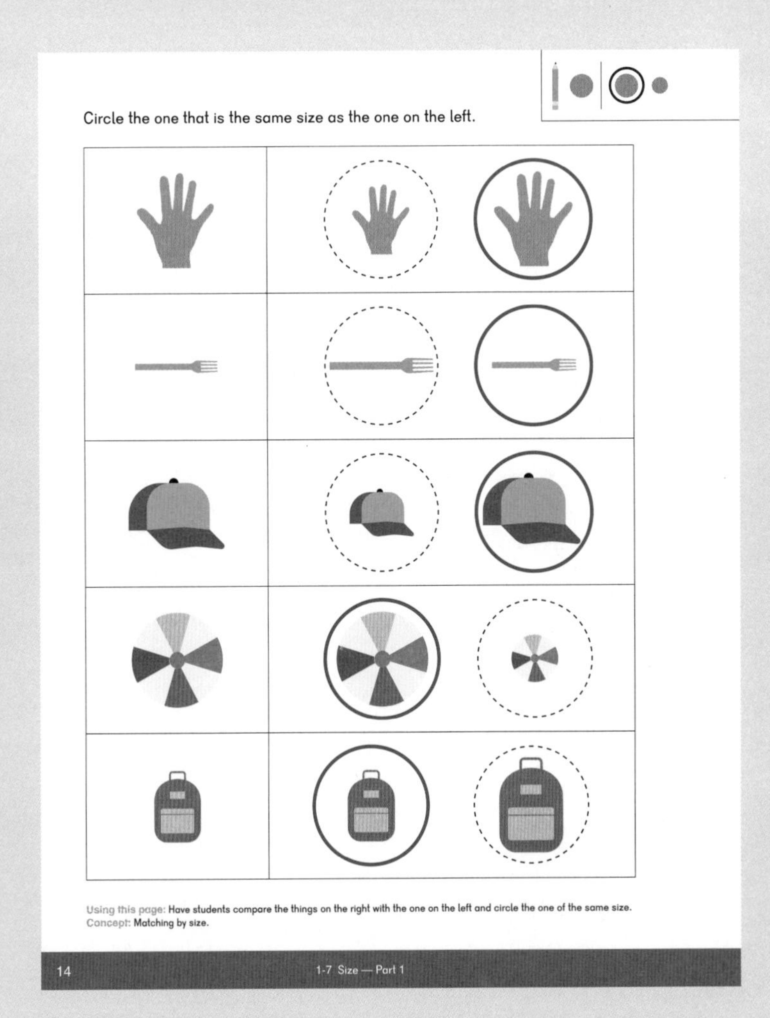

Exercise 8 • pages 15 – 16

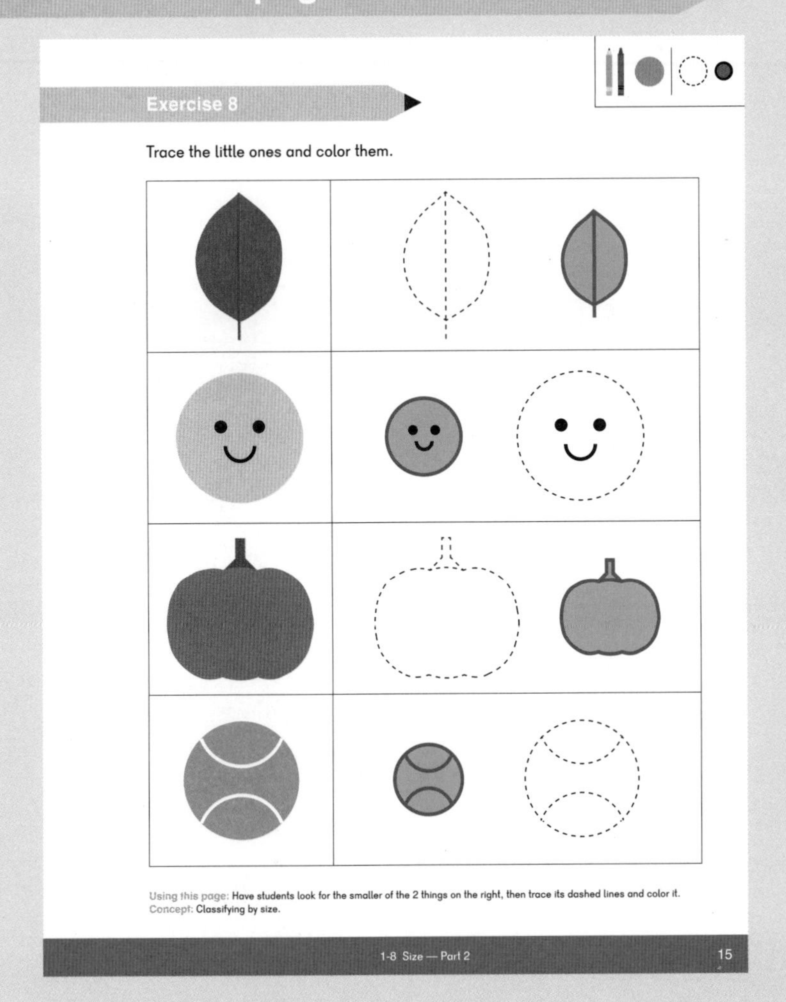

Exercise 9 • pages 17 – 18

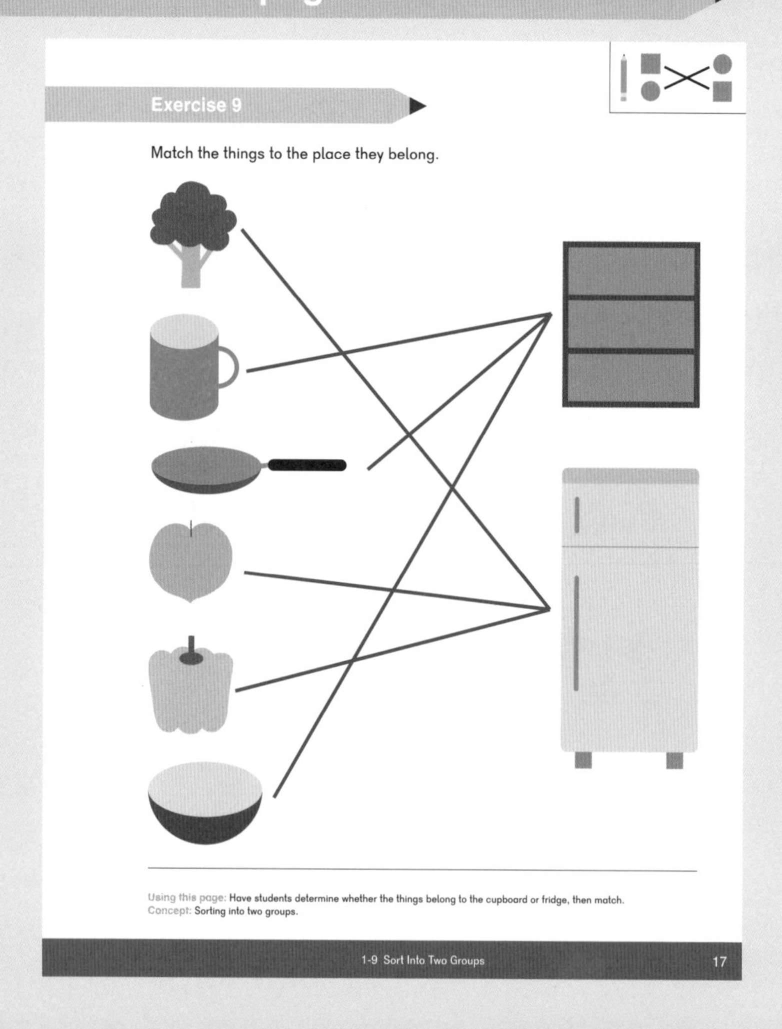

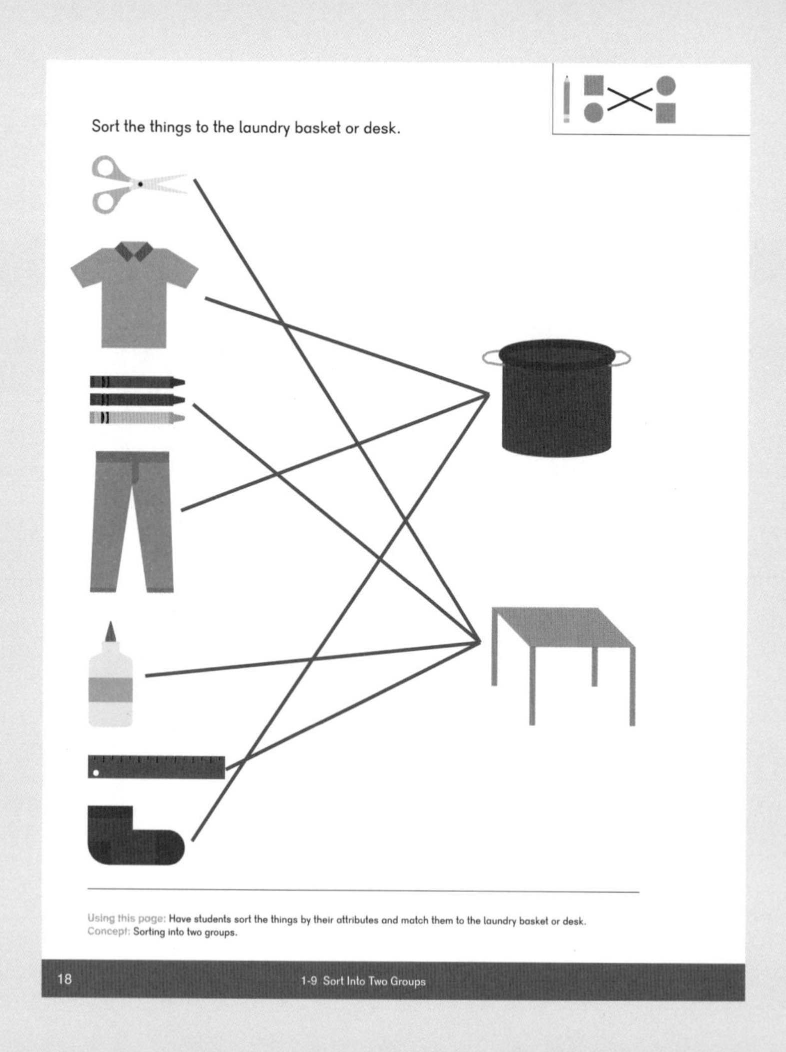

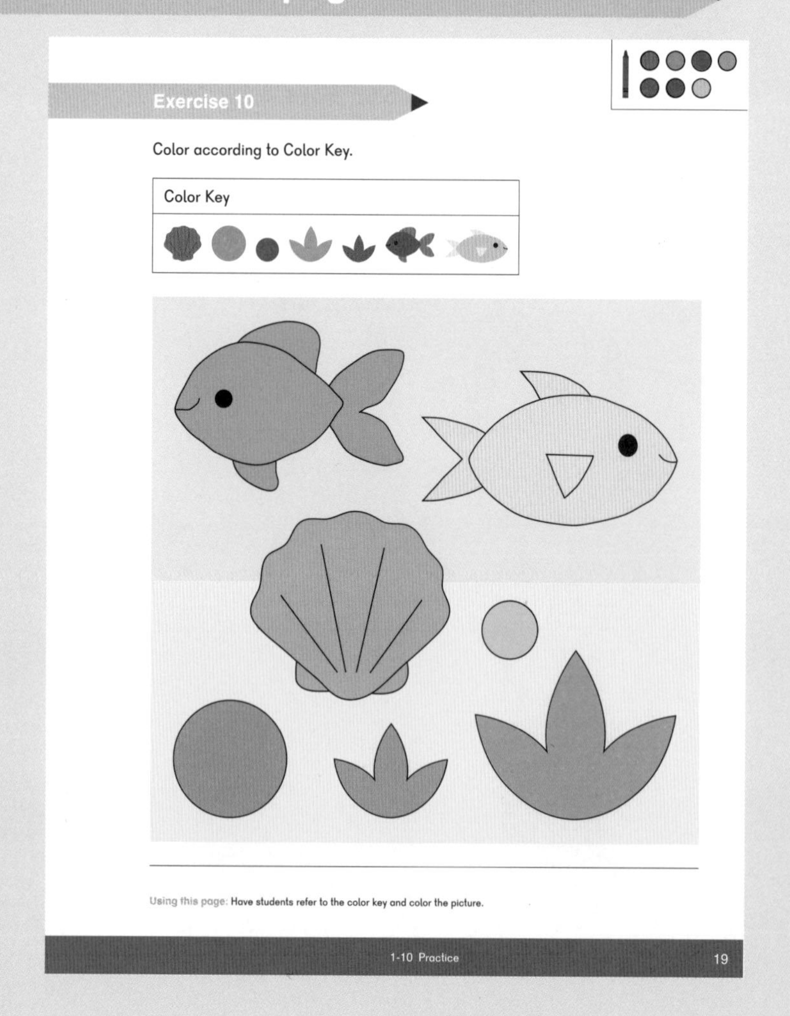

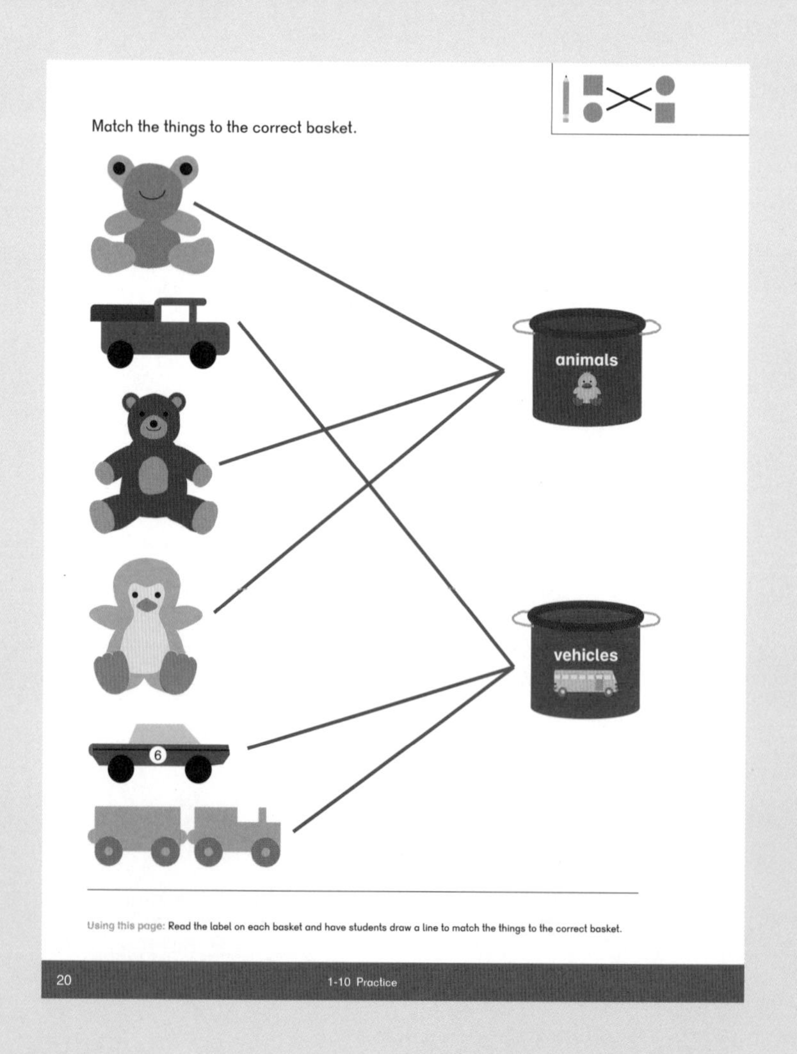

Suggested number of class periods: 5 – 6

Lesson		Page	Resources	Objectives
	Chapter Opener	p. 35	TB: p. 23	
1	Big and Small	p. 36	TB: p. 24 WB: p. 21	Compare objects by size.
2	Long and Short	p. 38	TB: p. 26 WB: p. 23	Compare objects by length.
3	Tall and Short	p. 40	TB: p. 28 WB: p. 25	Compare objects by height.
4	Heavy and Light	p. 42	TB: p. 30 WB: p. 27	Compare objects by weight.
5	Practice	p. 44	TB: p. 32 WB: p. 29	Practice concepts introduced in this chapter.
	Workbook Solutions	p. 46		

Chapter Vocabulary

- Compare
- Size
- Bigger than
- Biggest
- Smaller than
- Smallest
- Larger than
- Largest
- Length
- Short
- Shorter than
- Shortest
- Long
- Longer than
- Longest
- Height
- Tall
- Taller than
- Tallest
- Weight
- Heavy
- Heavier than
- Heaviest
- Light
- Lighter than
- Lightest
- Balance scales

Notes

In this chapter, students will begin to compare objects by size, length (including height), and weight.

All measurement done in Pre-K will be done by directly comparing the items being measured. In Kindergarten, students will begin to use non-standard units to measure. In Grade 1, students will continue to use non-standard units to measure and then compare the numerical value of the measurements. In Grade 2, they will learn about standard units of measurement.

In the previous chapter, students simply compared two items according to whether they were big or small. Here, students will compare relative size. When comparing only two objects, the comparative term, ending in "er," is used, and one object is being compared relative to another. So when presented simply with two objects, a student can be asked which is longer. However, if there are more than two objects, one object is not simply "longer," it is "longer than" some other object (and could be shorter than a third object), and students should include "than" to indicate which object it is being compared to.

Key Points

In *Classroom Instruction that Works*[1], the authors identify nine instructional strategies that improve student performance. Of the nine, the ability to think comparatively had the greatest effect on success.

Encourage use of math vocabulary. Correctly describing aspects of objects and comparing objects using math vocabulary helps students organize their thinking and clarify their reasoning to others.

Height is vertical length. In this chapter, length is explored in **Lesson 2: Long and Short** and height is explored in **Lesson 3: Tall and Short.**

---

[1] *Marzano, R. J., Pickering, D., & Pollock, J. E. (2001). Classroom Instruction That Works: Research-Based Strategies for Increasing Student Achievement. Alexandria, VA: Association for Supervision and Curriculum Development*

Materials

- Paper plates – different sizes
- Pieces of yarn/string – different lengths
- Pieces of colored paper – different colors, cut to different sizes
- Drawing supplies
- Objects that are similar but different sizes, such as:
    - 6 similar blocks
    - 6 pieces of paper the same color
- Tubs of linking cubes, 1 per pair of students
- Small bags – to hold objects and fruits
- Small objects to compare weight – e.g. marbles, pattern blocks, feathers, centimeter cubes, and toy cars
- Small fruits – e.g. tangerines, 1 per pair of students
- Balance scales

Note: Materials for Activities will be listed in detail in each lesson.

Blackline Masters

- None

Storybooks

- *Goldilocks and the Three Bears*
- *Lengthy, the Long Long Dog* by Syd Hoff
- *Steam Train, Dream Train* by Sherri Duskey Rinker

Optional Snacks

- Crackers (varying sizes)
- Celery sticks
- Apple slices
- Tangerines or other small, self-wrapped fruit
- Popcorn
- Carrot sticks

Letters Home

- Chapter 2 Letter

Notes

Lesson Materials

- Paper plates – different sizes
- Pieces of yarn/string – different lengths
- Pieces of colored paper – different colors, cut to different sizes

Chapter 2

Compare Objects

23

23

Explore

Give each student either a paper plate, piece of yarn, or piece of paper. Have them find a partner who has an object like theirs, then sit down together.

Have pairs discuss their objects using the phrases, "are the same because," "are different because," etc. Focus student attention to the whole group again. Ask students to share their discussions with the group.

Learn

Have students look at the illustration and ask them which things are the same and which are different, and explain their reasoning. Some things students might notice:

- The pets are almost the same because they all are covered in fur, have four legs, etc.
- The four pets are different because they are all different sizes.
- The cats are different because they are different colors.
- The dogs are almost the same because they are the same color.
- The dogs are different because one is big and one is small.
- The size and color of the yarn balls are not the same.

Extend Explore

Same and Different Drawing: Have each student draw a picture comparing himself or herself to a favorite superhero or character in a book. If possible, have each student self-narrate the details of the comparison into an audio recorder using the phrases "are the same because," "are not the same because," "are almost the same because," and "are different because."

Materials: Art paper, drawing supplies, recording device

Objective

- Compare objects by size.

Lesson Materials

- Objects that are similar but different sizes, such as:
  - 6 similar blocks
  - 6 pieces of paper the same color
- *Goldilocks and the Three Bears*
- Optional snack: Crackers of different sizes

24

Explore

Give each student one item from the materials list. Tell them to find other students who have objects like theirs, then sit down together in a group.

Focus student attention to the whole group again. Ask students to say why they sat together. If there is more than one group holding the same type of object, have groups combine.

Introduce the terms "compare," "bigger than," and "smaller than."

Have students talk in their groups to compare objects by saying whose is bigger than or smaller than someone else's.

Introduce the terms "biggest" and "smallest."

Have students in each group decide which of their objects is biggest and smallest.

Show five different sized pieces of paper. Help students realize that an object can be smaller than one object and at the same time bigger than a different object.

Read *Goldilocks and the Three Bears* to the students again. Focus on the pictures in the story and have students compare the sizes of the characters, beds, chairs, bowls, and spoons.

Learn

Discuss the illustration on page 24. Have students identify by color the chair, bowl, and spoon that each of the bears would use. Students may also look at textbook page 1 and compare the sizes of the objects on that page.

Have students look at Alex's toys on page 25 and discuss which, in each row, is smallest and which is biggest. Have them complete the task.

Whole Group Play

Catch: Have students play catch with or roll the different sized balls, comparing the sizes of the balls as they play.

Materials: 3 different sized balls

Comparing Sizes Simon Says: The usual Simon Says rules apply except that Simon will be telling students to do things which involve comparing sizes. For example:

- Put your smallest finger on your nose.
- Sit on the biggest rug.
- Stand next to something smaller than you.

Small Group Center Play

Sort: Similar objects of three different sizes into containers for big, bigger, and biggest.

Blocks: Build with blocks of different sizes.

Dress-Up: Include clothes in several sizes.

Art Without a Paintbrush: Have students dip various sized sponges into red, blue, yellow, green, and orange paint and transfer the paint to art paper. Have them compare the sizes of the sponges as they paint.

Exercise 1 • page 21

Extend Play

What Else Could Simon Say?: Have students help think of things for Simon to say using size comparative language.

Objective

- Compare objects by length.

Lesson Materials

- Tubs of linking cubes, 1 per pair of students
- Crayons
- Optional snack: Celery sticks of different lengths

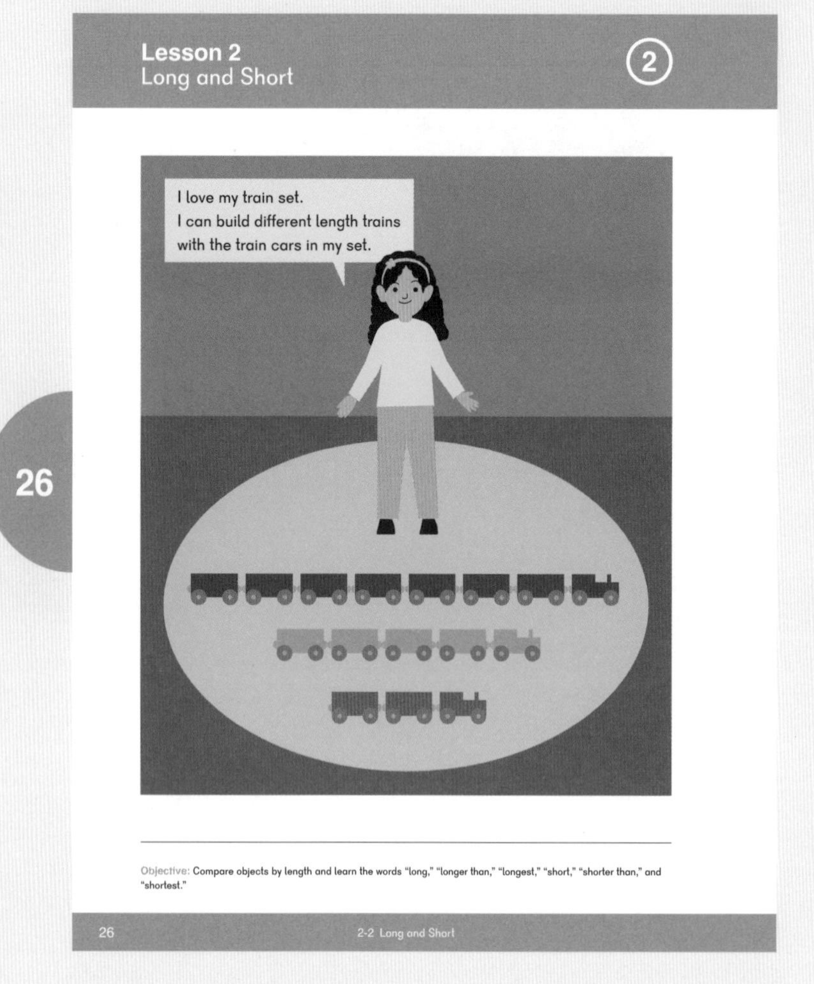

Explore

Ask students how they learned to compare objects in **Lesson 1: Big and Small**. Have them quickly point at something that is bigger than their nose, smaller than the flag, etc.

Provide each pair of students a tub of linking cubes and allow them to explore as they wish.

Ask them to build trains[1] by attaching several cubes together. Introduce the terms "short," "shorter than," "shortest," "long," "longer than," and "longest."

Demonstrate lining up objects at the same starting point to compare them by length, such as placing the end of each object against one line on the carpet.

Have students compare the lengths of their trains with that of their friends using comparative language. Remind them to have their trains line up at a common start point.

Learn

Have students compare the lengths of Sofia's trains on page 26. Ask them what is different about the way Sofia's trains are arranged from the way they arranged their trains to compare lengths. Does this difference make it easier or more challenging to compare the lengths? Why?

On page 27, read Sofia's directions aloud and have them cross out the longest pencil and circle the shortest pencil in the first group. Then have them complete the task.

---

[1]A train of linking cubes is a set of linked cubes in a straight line.

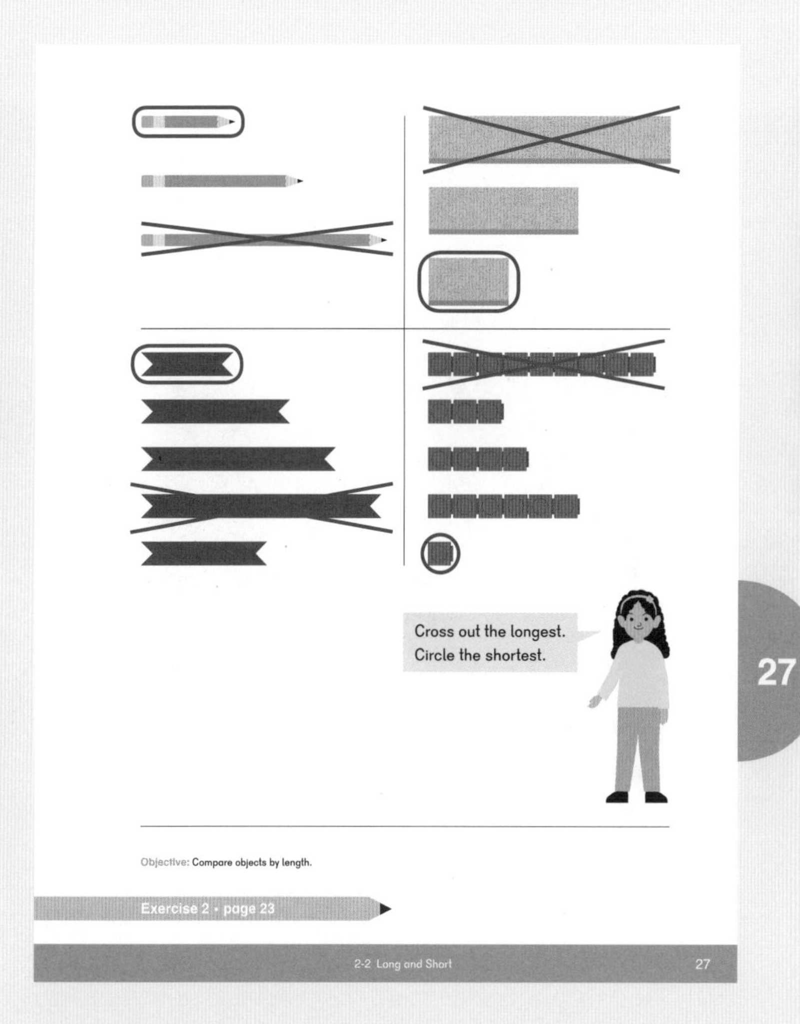

Whole Group Play

How Long Is Our Choo-Choo?: Put students in three groups, with a different number of students in each. The students will pretend to be trains while comparing the lengths of their trains.

Reading Time: Read *Lengthy, the Long Long Dog* to the students. Have them take turns acting out parts of the story.

Materials: *Lengthy, the Long Long Dog* by Syd Hoff

Small Group Center Play

Sort: Provide different lengths of yarn for students to sort by length.

Blocks: Have students build long objects.

Dress-up: Include clothes in several different sizes.

Kitchen: Provide multiple sizes of dishes and bowls for students to use.

Art Without a Paintbrush: Have students dip various lengths of yarn or string into red, blue, yellow, green, and orange paint and transfer the paint to art paper.

27

Exercise 2 • page 23

Extend Play

Reading Time: Read the book *Steam Train, Dream Train* to students. Challenge students to find classroom materials to build a train track, such as craft sticks. Then have them make a small train using a small cylinder as the base.

Materials: *Steam Train, Dream Train* by Sherri Duskey Rinker, craft sticks, small cylinders such as coin wrappers

Lesson 3 Tall and Short

Objective

- Compare objects by height.

Lesson Materials

- Optional snack: Apple slices cut to different lengths

Explore

Discuss the ways that students have learned to compare objects so far. Have them quickly point to something that is longer than a pencil you hold up, shorter than the windowsill, etc.

Ask a student to stand next to you. Have other students say which of you is tall and which is short.

Ask a third student to stand with you. Introduce the terms “taller than,” “tallest,” “shorter than,” and “shortest.” Ask students to compare your height to the heights of the two standing students using the new vocabulary.

Be sure that students understand that a person or object can be taller than one thing and shorter than another.

Put students into groups of three. Have each group of students arrange themselves by height.

Ask the students in each group which of them is tall, taller, and tallest. Stand next to the tallest student in each group and ask them, in that case, which of them is short, shorter, and shortest.

Learn

Have students look at the three friends and compare their heights.

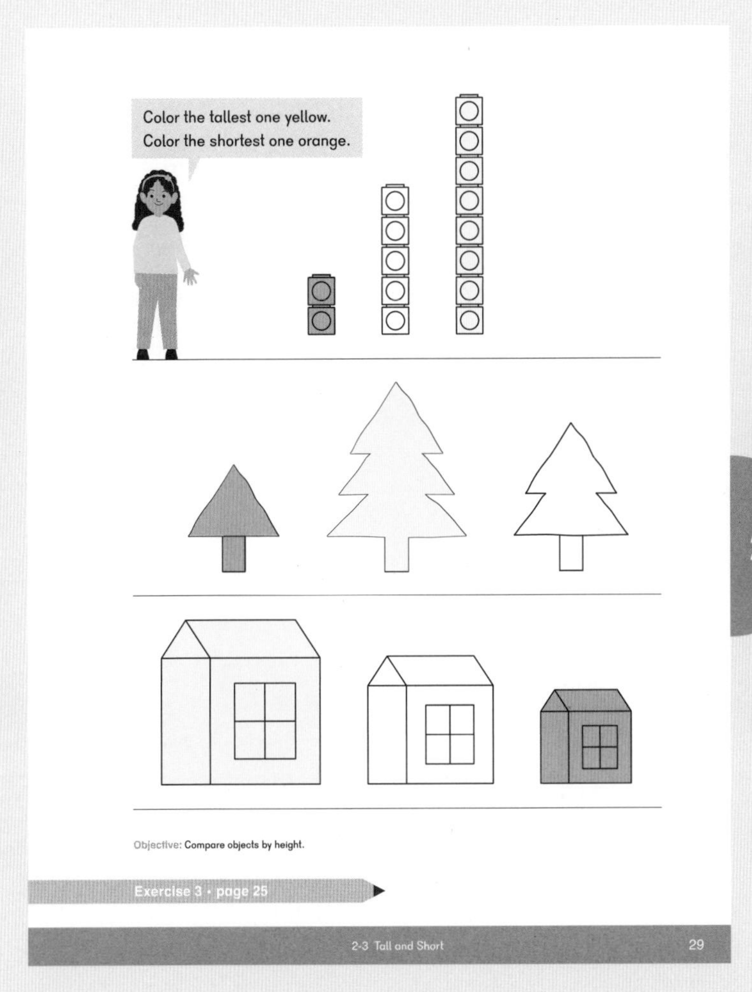

Read Sofia's directions aloud. Have students say which of the linking cube towers is the tallest and which is the shortest. Then have them complete the task.

Whole Group Play

Whose Toy is the Shortest?: Have each student bring a stuffed toy to class. Put all of the toys in the middle of the floor and have students arrange them by height. Whoever brought the shortest toy wins.

Small Group Center Play

Sort: Similar objects of three different heights into containers for short, shorter, and shortest.

Blocks: Have students build structures of different heights with blocks.

Dress-up: Include clothes in several different sizes.

Draw that Structure: Have students draw their structures created at the **Blocks** center.

29

Exercise 3 • page 25

Extend Play

Spaghetti Structures: Challenge students to build the tallest free-standing marshmallow and spaghetti sculpture they can in five minutes.

Materials: Large and small marshmallows, uncooked spaghetti

Objective

- Compare objects by weight.

Lesson Materials

- Small bags to hold objects and fruits
- Small objects to compare weight – e.g. marbles, pattern blocks, feathers, centimeter cubes, and toy cars
- Small fruits – e.g. tangerines, 1 per pair of students
- Balance Scales
- Crayons
- Optional snack: Fruit used in the lesson

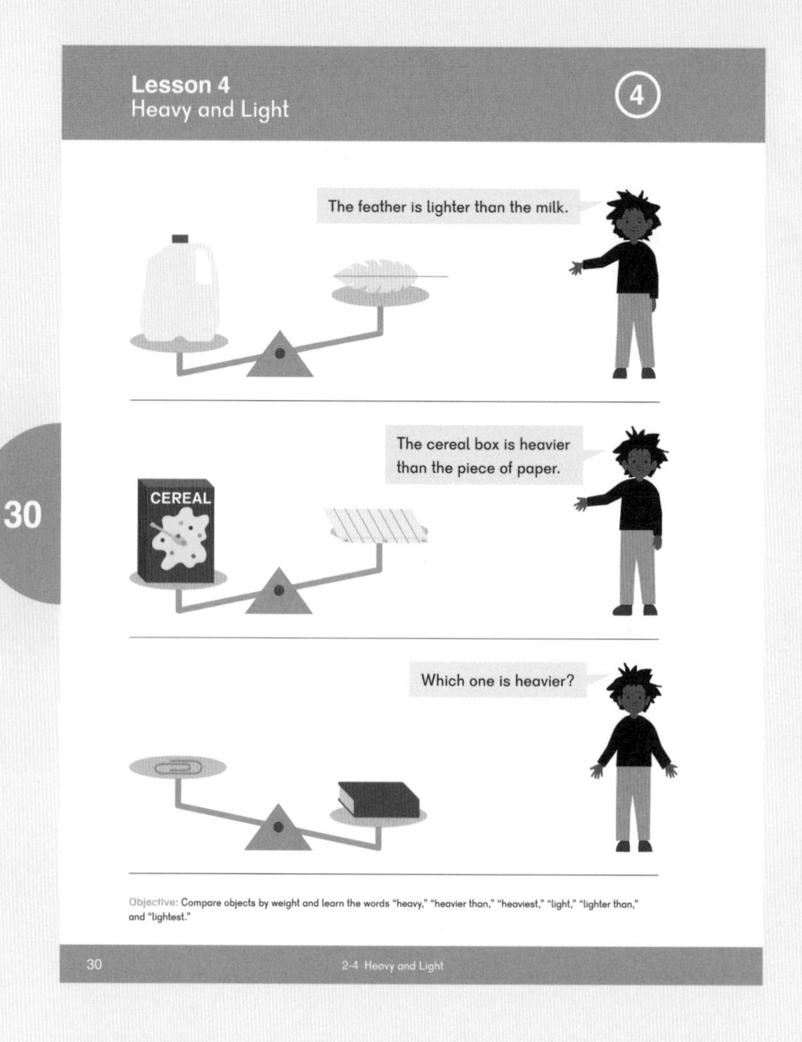

Explore

Have students name different ways to compare objects. Have them quickly point to something that is bigger than a lunch box, longer than a crayon you hold up, shorter than a bookcase, etc.

Prior to this lesson, prepare small bags with one of each item listed in **Lesson Materials.** Give each pair of students a bag of materials.

Have them remove the fruit and talk about whether they think the fruit is heavy or light with their partners.

Name one other object in their bags. Have them remove that object and hold the fruit in one hand and the object in the other. Introduce the words "heavier" and "lighter." Ask them which is heavier, the fruit or the other object.

Repeat with the other objects, alternating between asking which object is lighter and which is heavier. If students cannot tell which object is heavier or lighter just by holding the objects, ask them how they could find out for sure which object is heavier.

After all of the objects in the bag have been compared, choose three objects of different weights. Introduce the terms "heavier than," "heaviest," "lighter than," and "lightest." Have students hold the objects in their hands, two at a time, and compare their weights.

Demonstrate using balance scales to compare the weights of the three objects. Tell students that one of the objects is heavier than another, but lighter than the third.

Learn

Have students look at the objects on page 30 and say what they are. Ask them which is heavier, a gallon of milk or a feather. Read Dion's comments and discuss the appearance of the objects on the scales.

Have them answer Dion's question at the bottom of the page.

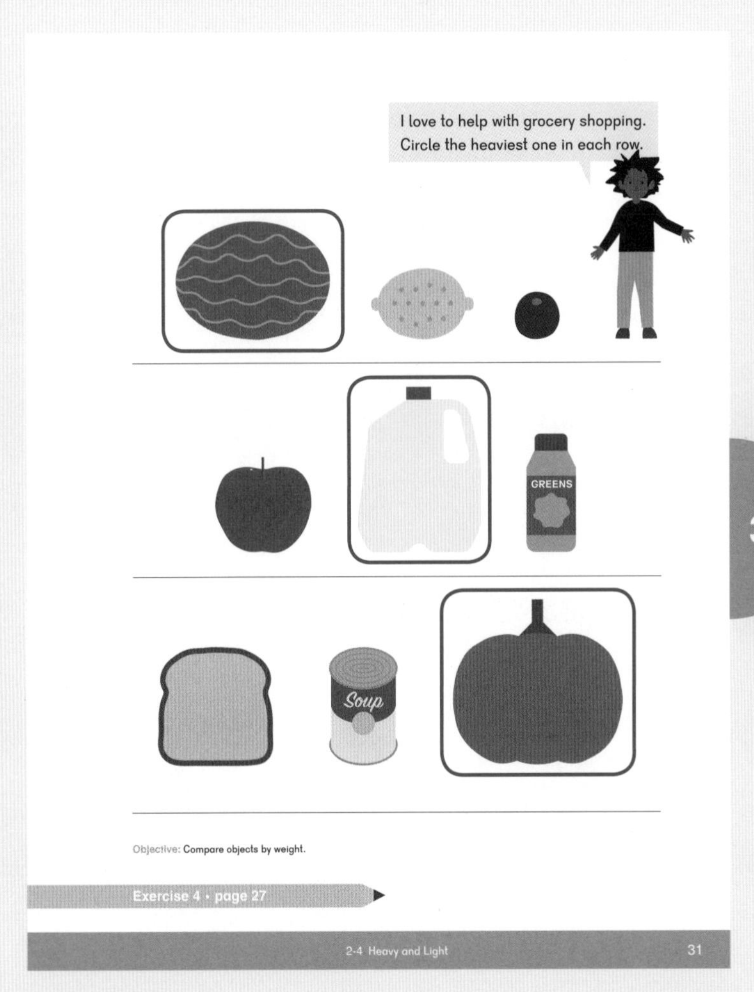

31

Have students identify the food items on page 31. Read Dion's directions aloud and have them complete the task.

Whole Group Play

Nature Walk: Take a nature walk and talk about weight. Ask students questions about the weights of objects, for example, "Which do you think is heavier, a blade of grass or a rock?," "Which do you think is lighter, a pinecone or a tree branch?," and, "Can you point to something that you think is heavier than a twig?" Encourage discussion.

Small Group Center Play

Sort: Objects by weight.

Blocks: Build with blocks of different weights.

Dress-up: Include clothes of different weights, such as a light T-shirt and a heavy jacket.

Art Without a Paintbrush: Have students dip feathers or cotton balls into red, blue, yellow, green, and orange paint and transfer the paint to art paper.

Balance Scales: Set up several balance scales and small objects to weigh. Have students compare the weights of objects.

Exercise 4 • page 27

Extend Explore

Lightest to Heaviest: Have students order the objects in their bin from lightest to heaviest. After the objects have been ordered, give each student one more object, different from any of the others, and have him or her determine where in the order that object should go.

Materials: Bins containing 6 – 7 objects of varying weight for each student or group of students

Objective

- Practice concepts introduced in this chapter.

Lesson Materials

- Optional snack: Popcorn and various lengths of carrot sticks

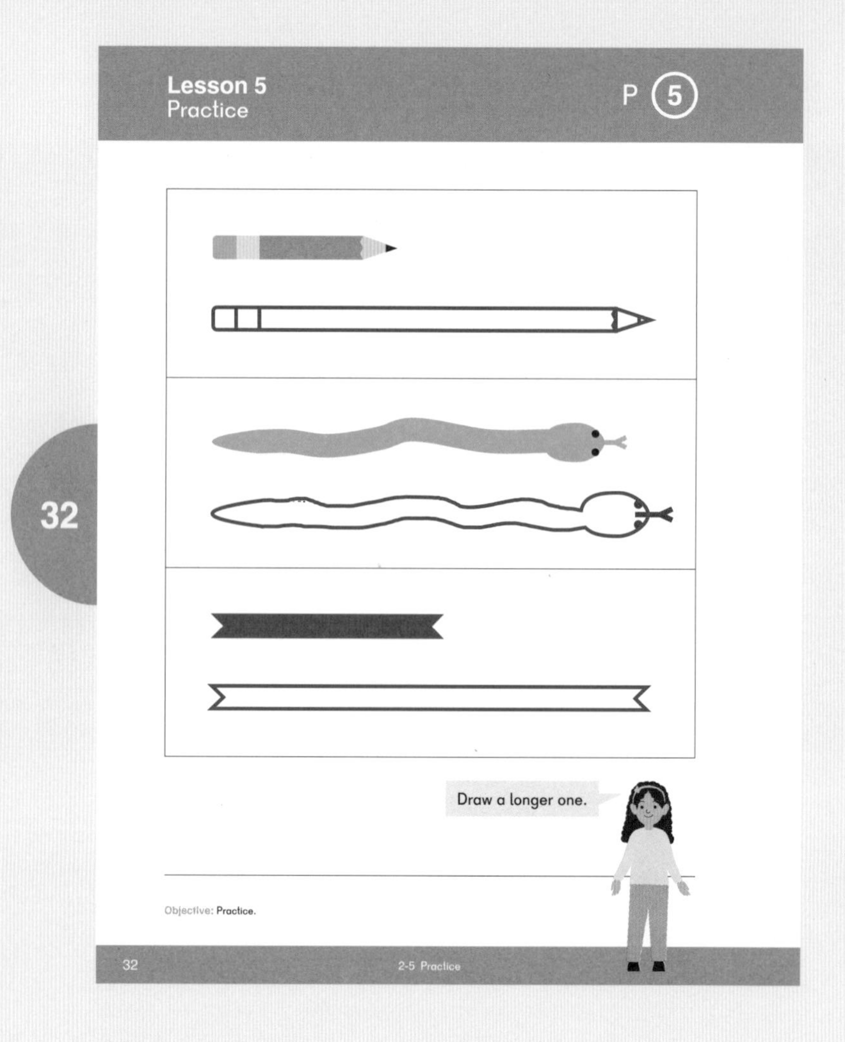

For the **Practice,** read the directions and speech bubbles on each page and have students complete the tasks.

Whole Group Play

Parachute Popcorn: Put a dozen cotton balls in the middle of a play parachute. Have students each hold two of the parachute handles and move their arms up and down to have the cotton balls "pop." Next, put a dozen bean bags in the middle of the parachute and repeat the activity. Have students discuss the difference between trying to "pop" cotton balls and trying to "pop" bean bags. Ask them why the cotton balls are easier to "pop."

Small Group Center Play

Sort: Similar objects of three different lengths to be sorted into bins for short, shorter, and shortest.

Blocks: Build with blocks of different weights.

Dress-up: Include clothes of several different sizes and weights.

Art Without a Paintbrush: Have students dip various lengths of yarn or string into red, blue, yellow, green, and orange paint and transfer the paint to art paper.

Balance Scales: Set up several balance scales and small objects to weigh. Have students compare the weights of objects.

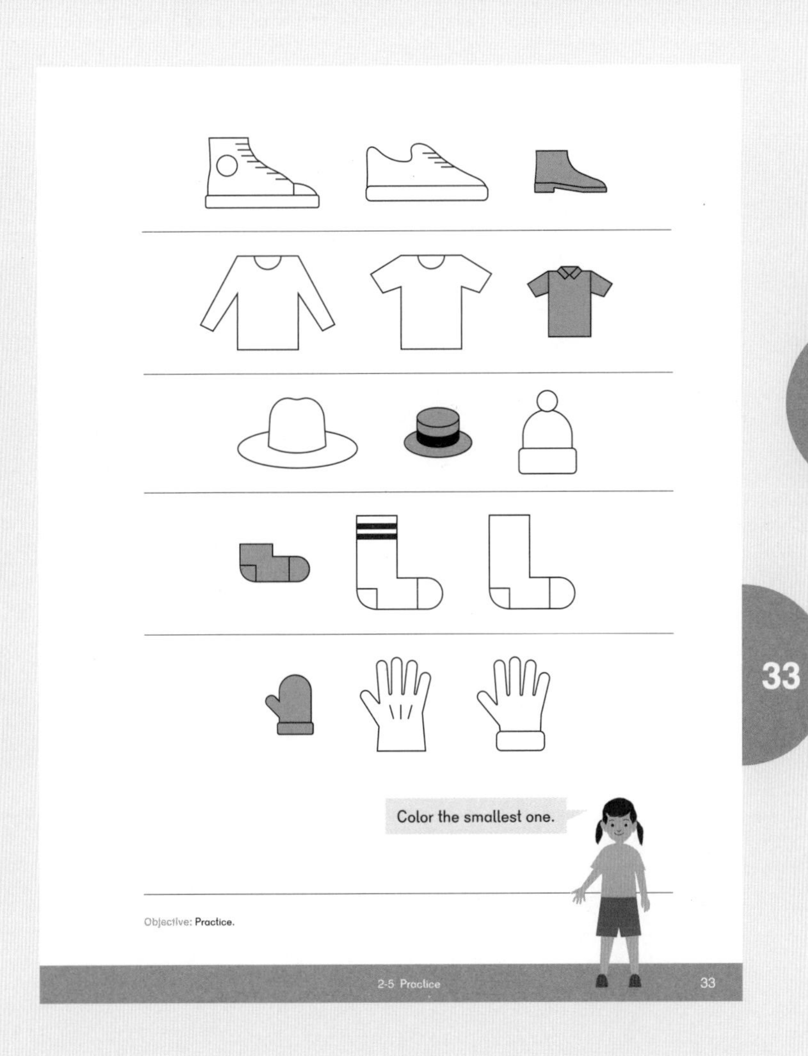

Exercise 5 • page 29

Extend Explore

Make it Balance: Have students use balance scales to figure out how many linking cubes weigh the same as various small toys.

Materials: A container of linking cubes and several small toys for each student, balance scales

Exercise 1 • pages 21 – 22

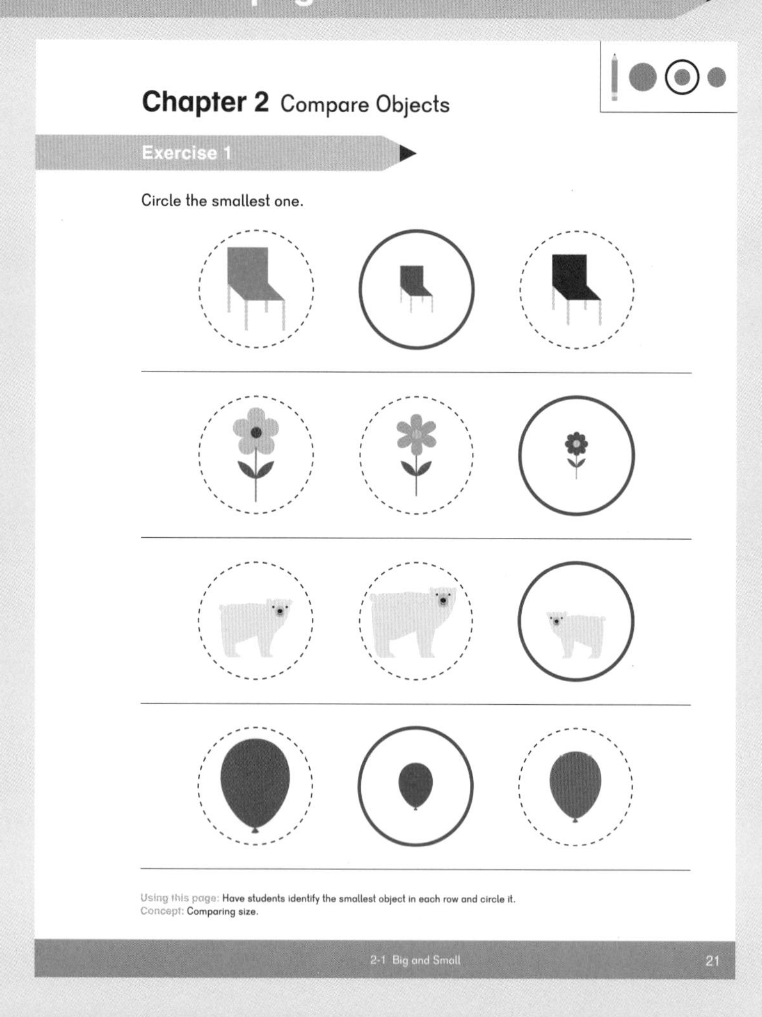

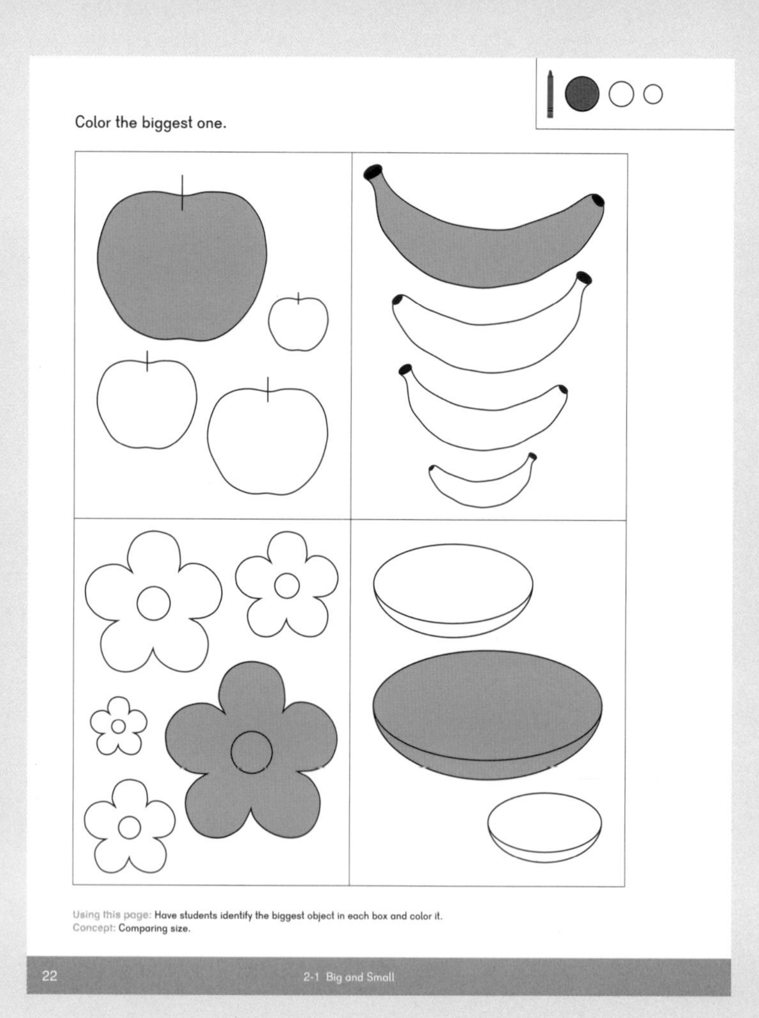

Exercise 2 • pages 23 – 24

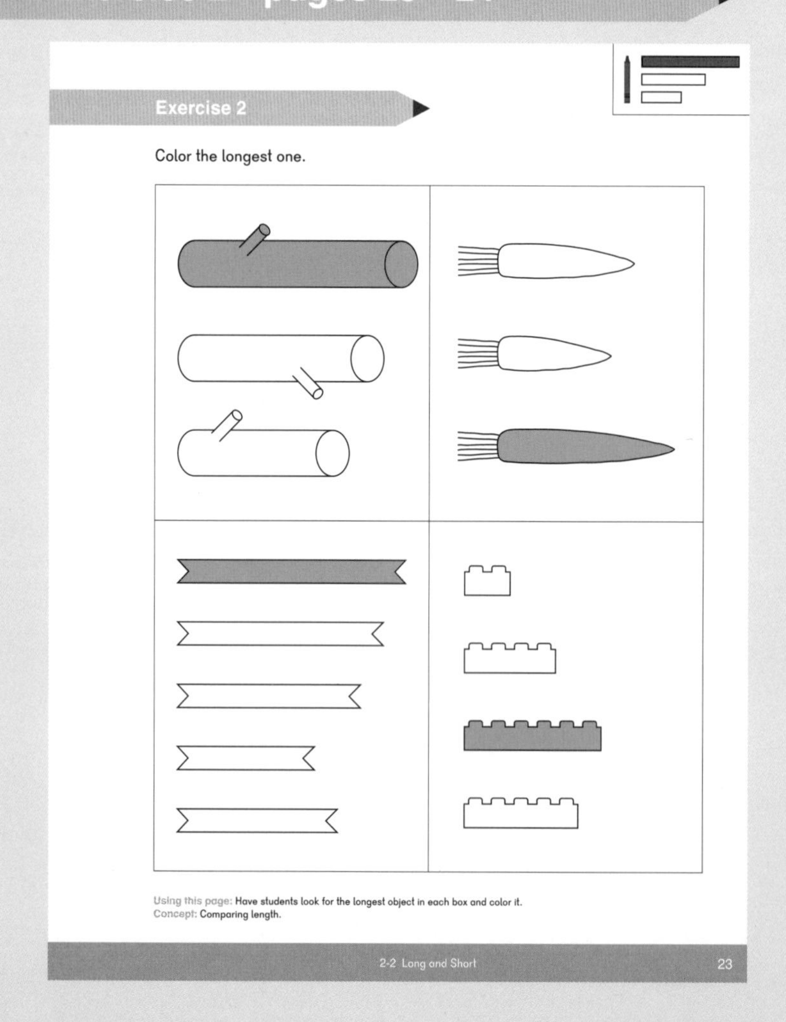

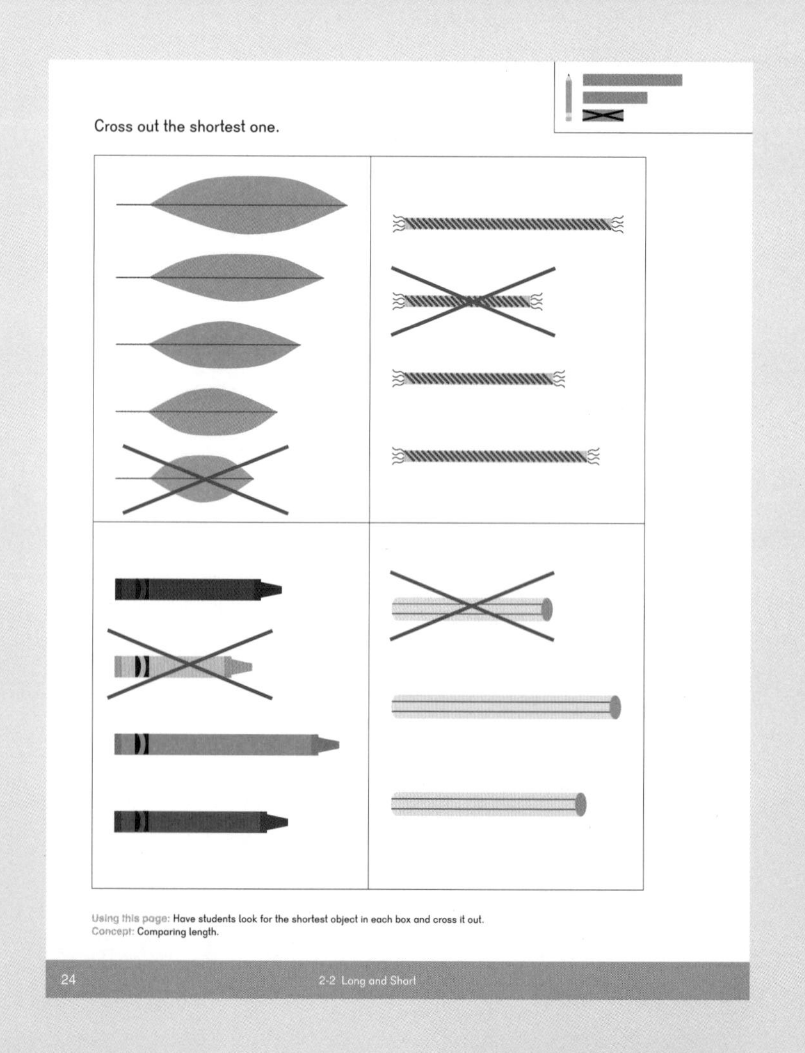

Exercise 3 • pages 25 – 26

Exercise 4 • pages 27 – 28

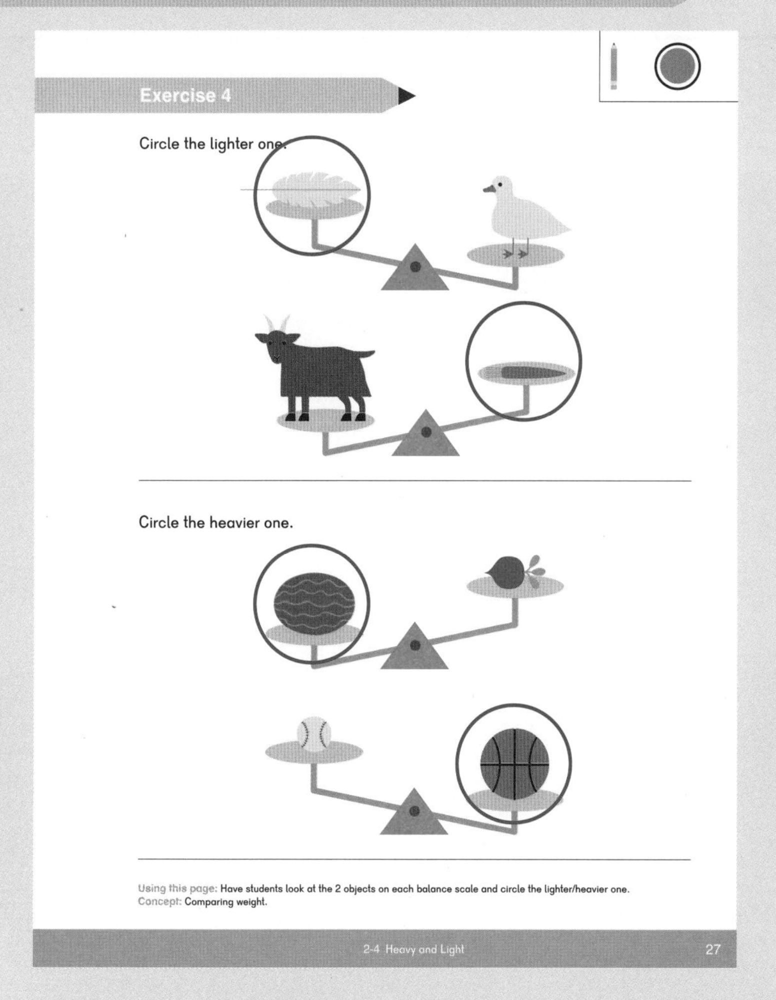

Exercise 5 • pages 29 – 30

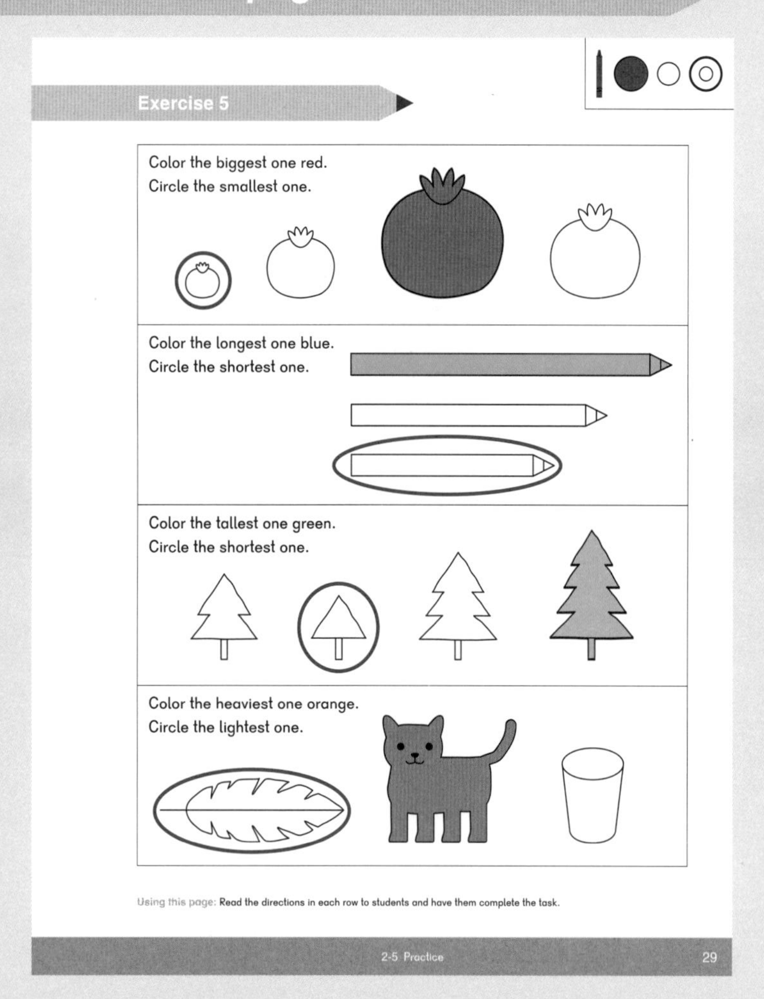

Suggested number of class periods: 4 – 5

Lesson		Page	Resources	Objectives
	Chapter Opener	p. 53	TB: p. 35	
1	Movement Patterns	p. 54	TB: p. 36 WB: p. 31	Create AB patterns using movement.
2	Sound Patterns	p. 56	TB: p. 38 WB: p. 33	Create AB, AAB, and ABB patterns using sound.
3	Create Patterns	p. 58	TB: p. 40 WB: p. 35	Create patterns. Recognize the colors black and white.
4	Practice	p. 61	TB: p. 43 WB: p. 37	Practice concepts introduced in this chapter.
	Workbook Solutions	p. 63		

Chapter Vocabulary

- Pattern
- Repeating
- Black
- White
- Movement
- Sound
- Create

Notes

Recognizing, extending, and creating patterns is a critical math concept for young children. Having this ability will help students predict what comes next, develop mental math strategies, skip count, and think algebraically.

Key Points

Young children recognize patterns in their daily lives. Recognizing patterns is among the early concepts suggested for ongoing success in the study of math. In Pre-K, students will look for which part of a pattern repeats itself and will focus on identifying and extending patterns. They will be introduced to simple patterns, including AB, AAB, and ABB. Students are not expected to name the types of patterns.

Materials

- Small bags of linking cubes – 2 different colors, several of each color, 1 bag per student
- Two types of objects – e.g. forks and spoons
- Pattern blocks
- 2 types of noise makers – e.g., bells, drums, spoons, kazoos
- Small bag of linking cubes – 3 different colors, 15 cubes per bag
- Crayons

Note: Materials for Activities will be listed in detail in each lesson.

Blackline Masters

- None

Storybooks

- *Bees, Snails & Peacock Tails: Patterns & Shapes* by Betsy Franco
- *Hickory Dickory Dock*

Optional Snacks

- Apple slices
- Cheese sticks
- Pretzel sticks
- Strawberry slices
- Banana slices
- Crackers in a variety of shapes

Letters Home

- Chapter 3 Letter

Notes

Lesson Materials

- Small bags of linking cubes – 2 different colors, several of each color, 1 bag per student
- Two types of objects – e.g. forks and spoons

Explore

Prior to this lesson, create small bags of linking cubes (two different colors). Prepare enough so each student will receive one bag.

Lay two types of objects in front of you for students to see, naming the objects as you lay each one down. For example, you might say, "Fork, spoon, fork, spoon, fork, spoon." Point to one object at a time and have students name them as you do. Ask them to say what object would come next in the pattern you are showing in the example, and they should say, "Fork."

Give each student a bag of linking cubes and have them create a repeating color pattern such as, "red, blue, red, blue, red, blue, ..." etc.

Have students tell a partner about their patterns by saying aloud the colors of the linking cubes.

Learn

Have students look at page 35 and say what is the same and what is different about the flowers in the two flower beds. Encourage them to describe the flowers aloud by color in the top flower bed and by height in the second flower bed.

Chapter 3

Patterns

35

Extend Explore

Explore Patterns: Allow students to play with pattern blocks. Notice which students use the blocks to create repeating patterns.

Materials: Pattern blocks

Objective

- Create AB patterns using movement.

Lesson Materials

- Optional snack: Apple slices and cheese sticks (gymnasts and dancers love this combination!)

Explore

Ask students to find something in the room that follows a repeating pattern. They may point to fabric in clothes, a poster on the wall, etc.

Demonstrate several body movements, such as putting hands on head, kicking a foot, and touching toes. Have students practice each movement with you several times. Model an AB pattern for them, naming each movement aloud as you make it. Ask students to create their own patterns using two movements.

After a few minutes, ask individual students to demonstrate the movement patterns they created and have all students practice the patterns shown.

Learn

Have students look at page 36 and discuss Emma's comments.

Ask students if any of them take gymnastics lessons. If so, have them lead the other students in some movements. Have students use two movements to create a movement pattern. Lead them in other AB pattern movements, naming aloud each movement as you make it.

Have students look at page 37 and identify the pattern on the floor (light brown, brown, light brown, brown, etc.). Discuss Alex's comments. Repeat the procedure from the previous page to have students create AB dance movement patterns.

Whole Group Play

High Kick Time: Have students line up in a row and put their arms over the shoulders of the students on either side. Play music and have students try to do high kicks: right leg, left leg, repeat.

Materials: Music with a clear beat

37

Exercise 1 • page 31

Extend Play

Drum Patterns: Using drums, encourage students to use their hands to tap the drums in AB, AAB, and ABB patterns.

Materials: Drums, either real or student-made

Small Group Center Play

- **Blocks:** Have students create an AB pattern by size of block: large, small, large, small, etc.
- **Dress-up:** Include clothes with clear patterns such as stripes.
- **Video Move:** Show a video of children dancing and have students dance.
- **Dot Art:** Have students dip cotton swabs into two colors of paint and create AB patterns on art paper.
- Allow students to explore the instruments freely.

Objective

- Create AB, AAB, and ABB patterns using sound.

Lesson Materials

- *Hickory Dickory Dock*
- 2 types of noise makers – e.g., bells, drums, spoons, kazoos
- Music
- Optional snack: Pretzel sticks (look like drum sticks)

Explore

Show students a quick movement pattern, such as kick, stomp, repeat. Ask them what movement comes next.

Lead students in an AB clapping pattern, such as: use hands to clap legs, clap hands together, repeat. Have students work in pairs. One student creates an AB sound pattern while the other copies it. Students then reverse roles.

Teach students to say, "Tick, tock, tick, tock," in rhythm, always saying "tick" louder than they say "tock." After they have become proficient at doing this without you, have them say those words as you read *Hickory Dickory Dock* to them, meeting their rhythm as you read the rhyme.

Learn

Have students look at page 38. Read Dion's directions. Then lead them in an AB clapping exercise to music.

Divide students into two groups. Provide the students in one group with one type of noise maker and the students in the other group a different type of noise maker. Allow them to play with their instruments for several minutes. Ask the students in each group to use their instruments to make sound patterns. Share the patterns created and discuss them.

Have students look at page 39 and name the instruments that Sofia and Alex are holding. Discuss patterns used in music.

39

Whole Group Play

Word Sound Patterns: Have students say animal sounds in an AAB pattern in their groups, such as, "Oink, oink, cluck, oink, oink, cluck," etc.

Musical Sound Patterns: Have students sing musical notes in an ABB pattern, such as high note, low note, low note, high note, etc.

Small Group Center Play

Patterns: Create an AB pattern by size of block: large, small, large, small, etc.

Dress-up: Include clothes with clear patterns such as stripes.

Video Move: Show a video of children dancing and have students dance.

Dot Art: Have students dip cotton swabs into two or three colors of paint and create AB or AAB patterns on art paper.

Music: Have students work in pairs to use two instruments to create AB or ABB patterns.

Exercise 2 • page 33

Extend Play

Clap Hands, Clap Legs: With students sitting cross-legged, teach them clapping patterns alternating clapping hands together then clapping hands on legs.

Objectives

- Create patterns.
- Recognize the colors black and white.

Lesson Materials

- Small bags of linking cubes – 3 different colors, 15 cubes per bag, 1 bag per student
- Crayons
- Optional snack: Strawberry slices and banana slices in a pattern on a stick

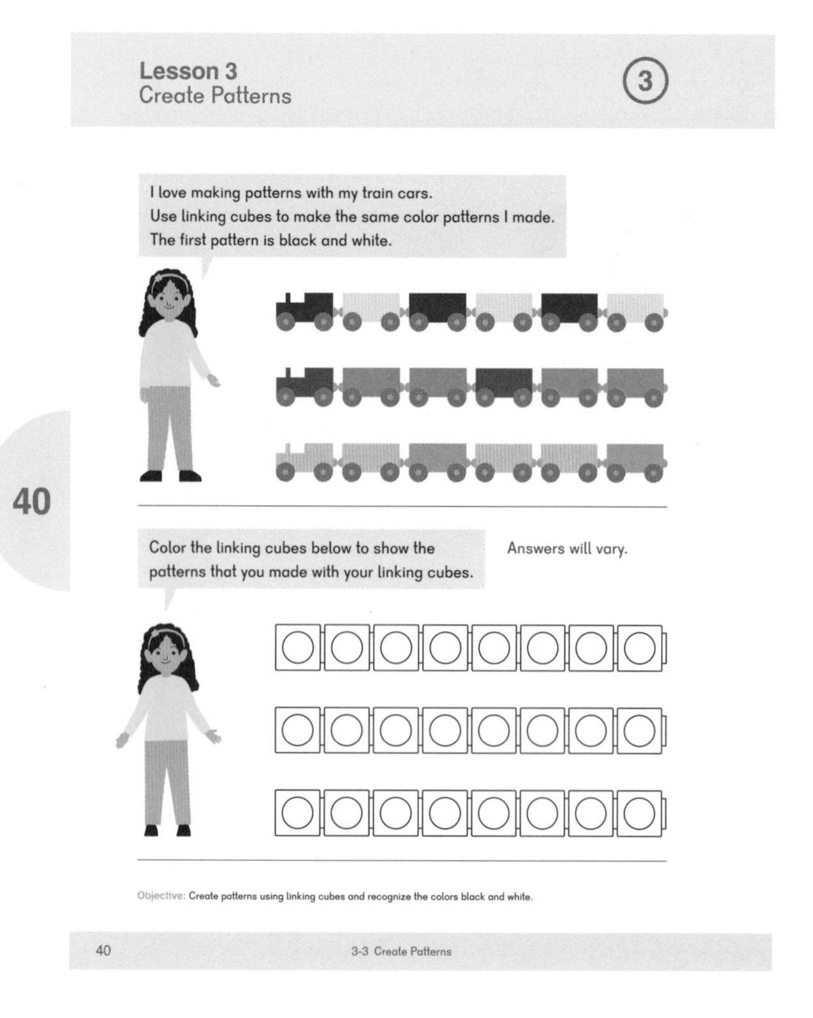

Explore

Prior to this lesson, create small bags containing linking cubes of three different colors, 15 cubes per bag, 1 bag per student. Ask them to create patterns of their choice. Some students may choose to use two colors. Others may use three. After a few minutes, ask other students what color linking cube would come next in some of the patterns. Students should say, for example, "Yellow, orange, yellow, orange, the next linking cube should be yellow!"

Introduce the colors black and white. Have students name objects in the room that are black or white.

Learn

Have students look at page 40 and discuss Sofia's comments. Name the color pattern in the first train from left to right with the students. Ask them what color car would come next.

Have students build one of the trains using linking cubes, then create other patterns with their linking cubes and color the cubes appropriately on the page.

Have students color page 41 appropriately after creating a string of beads during **Small Group Center Play**. This could be done on another day, if necessary.

Whole Group Play

Pattern Paper Chains: Have students create long paper chains using a pattern, and have them describe their patterns.

Materials: Paper strips of two or three different colors

Reading Time: Read *Bees, Snails, & Peacock Tails: Patterns & Shapes* to students. As you read, give them time to look for patterns on each page.

Materials: *Bees, Snails, & Peacock Tails: Patterns & Shapes* by Betsy Franco

Small Group Center Play

Patterns: Create a pattern of choice using blocks.

Dress-up: Include clothes with clear patterns such as stripes.

Kitchen: Use play kitchen items to create a pattern of choice.

Bead Necklaces: Provide string and beads of different colors. Have students create eight-bead necklaces using patterns of their choice. Then have them color their patterns on page 41.

Music: Have students work in pairs to use two instruments to create patterns of choice. If possible, record the musical patterns created.

Have students look at page 42 and identify the objects. Name the object pattern in the first row, "Maraca, maraca, flute, maraca, maraca, flute, maraca, maraca." Have students tell you which instrument comes next and have them circle it.

Ask the students what is different about the ribbons in the same row (stripes, dots, dots, etc.). Read Alex's direction and have them complete the task.

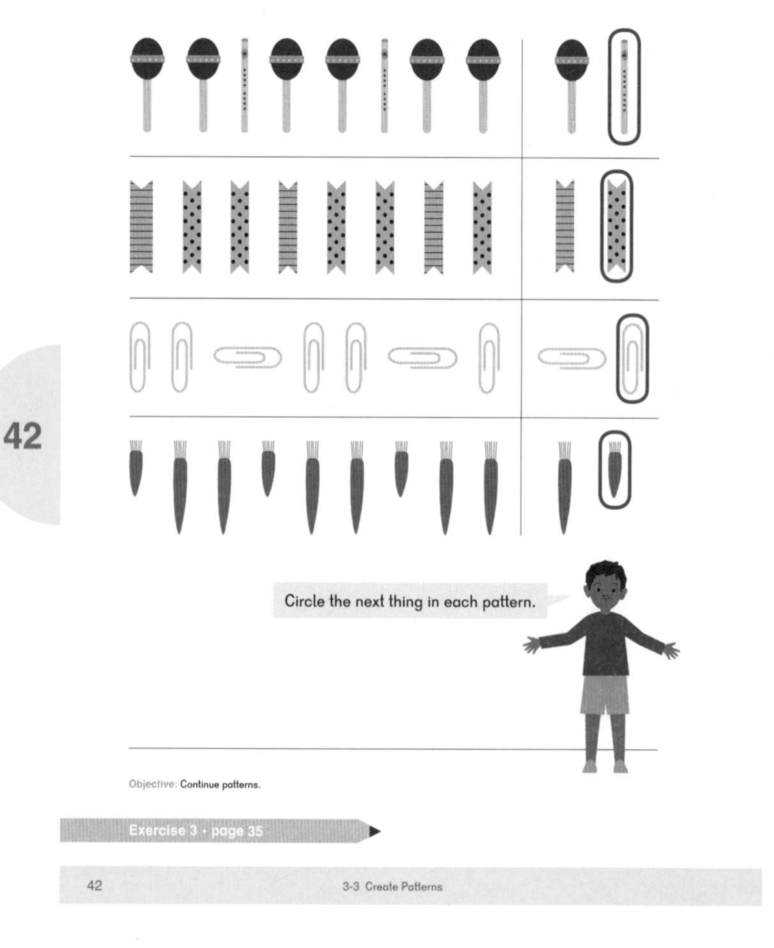

Exercise 3 • page 35

Extend Play

Discuss and Make Patterns: Before class, make paper chains using AB, AAB, ABB, and ABC patterns. Show them to students, and have them discuss similarities and differences between the different chains. Hand out paper strips to students and invite each of them create a paper chain using a pattern of choice. Have them trade chains with another student and continue the pattern.

Materials: Paper strips of two or three different colors

Objective

- Practice concepts introduced in this chapter.

Lesson Materials

- Optional snack: Crackers of different shapes

For the **Practice**, read the directions and speech bubbles on each page and have students complete the task.

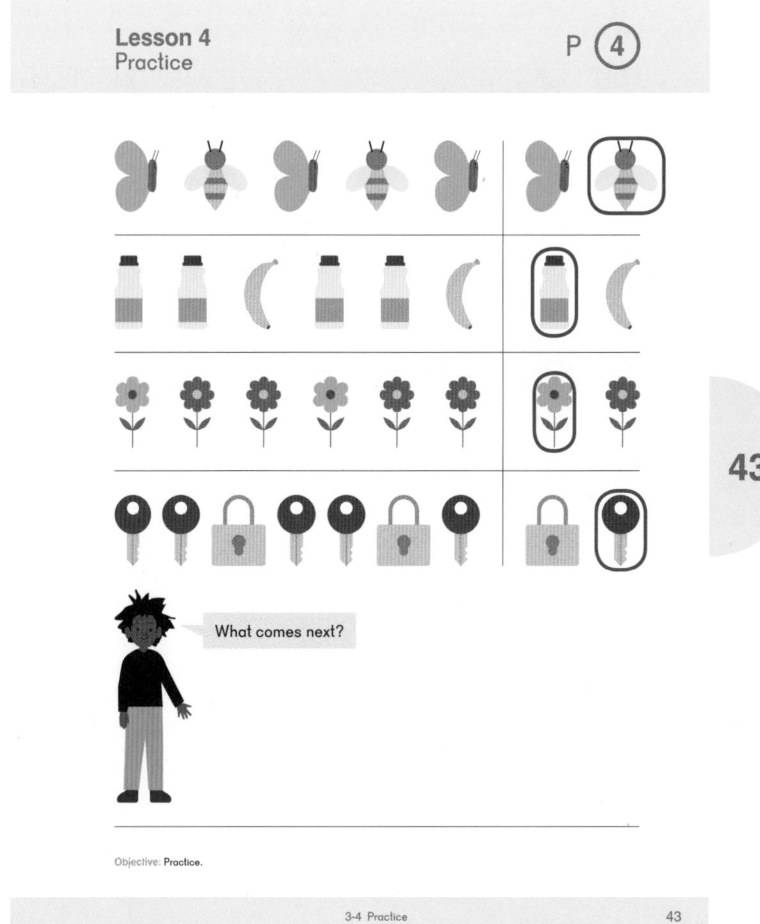

Whole Group Play

A Sailor Went to Sea: If time allows, teach students the nursery rhyme "A Sailor Went to Sea," and a clapping pattern that goes with it (VR).

A sailor went to sea, sea, sea
To see what he could see, see, see,
But all that he could see, see, see
Was the bottom of the deep blue sea, sea, sea

Two students sit on the floor facing each other. They then practice a clapping pattern to go along with the rhyme such as:

A – Clap their own hands together
Sai – Clap right hands
Lor – Clap hands together
Went – Clap left hands
To – Clap their own hands together
Sea, sea, sea – Clap both hands with partner three times

Repeat

Materials: A Sailor Went to Sea (VR)

Small Group Center Play

Patterns: Create a pattern of choice using blocks.

Dress-up: Include clothes with clear patterns such as stripes.

Kitchen: Use play kitchen items to create a pattern of choice.

Can You Make What I Made?: Create a pattern using pattern blocks. Put it in a box and tell students not to move the pattern blocks in the box. Provide pattern blocks and have students copy your pattern.

Music: Have students work in pairs to use two instruments to create sound patterns of choice. If possible, record the musical patterns created.

Exercise 4 • page 37

44

Answers will vary.

Color each tower using 2 different colors to create a pattern.

Objective: Practice.

Exercise 4 • page 37

44 3-4 Practice

Extend Play

The Bottom of the Sea: Have students describe orally, into a recording device if possible, what a sailor might see at the bottom of the sea. Encourage them to discuss what types of objects and/or technology might allow a sailor to see clearly underwater. If time allows, have them draw pictures of their descriptions. Suggest that students use a repeating pattern as part of their illustrations.

Materials: Recording device, art paper, drawing materials

Exercise 1 • pages 31 – 32

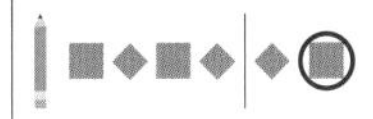

Chapter 3 Patterns

Exercise 1

Circle the one that comes next.

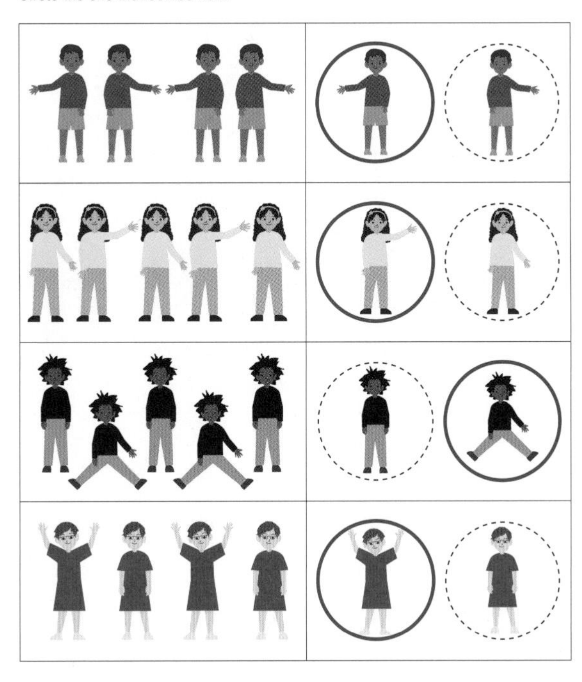

Using this page: Have students look at the movement in each row and trace the dashed line around the next movement.
Concept: Recognize AB patterns in movement.

3-1 Movement Patterns 31

Paste the one that comes next.

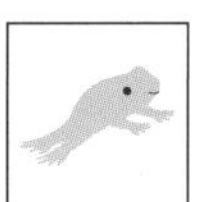

Before using this page: Pre-cut the pictures of the cat, frog, duck, and dog from page 121 for students to choose and paste.
Using this page: Have students look at the movement in each row and choose the correct picture to paste.

32 3-1 Movement Patterns

Exercise 2 • pages 33 – 34

Exercise 2

Color the one that comes next.

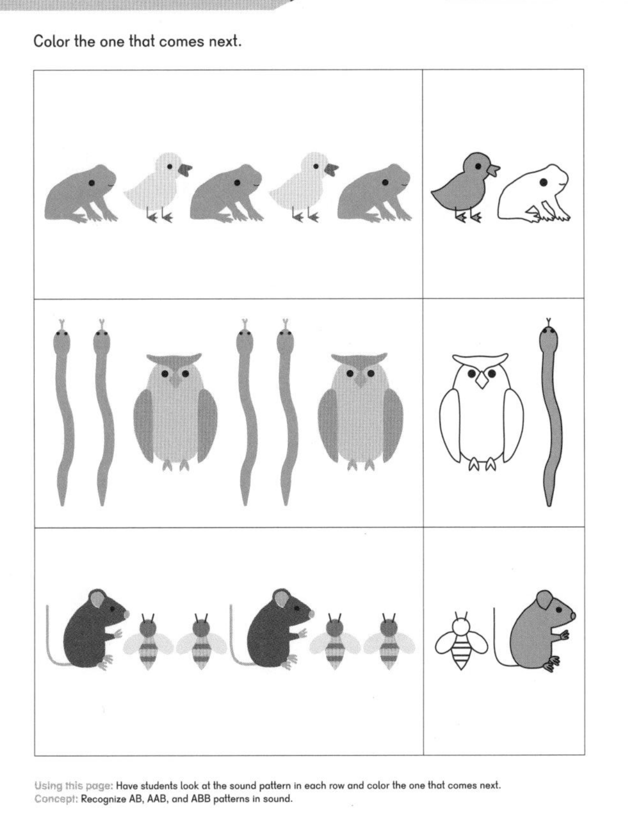

Using this page: Have students look at the sound pattern in each row and color the one that comes next.
Concept: Recognize AB, AAB, and ABB patterns in sound.

3-2 Sound Patterns 33

Circle the one that comes next.

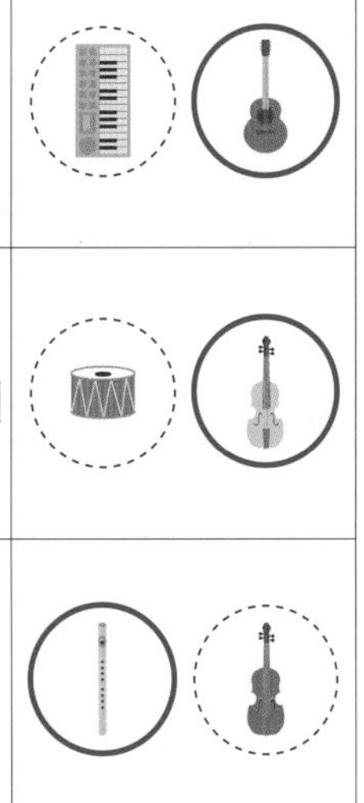

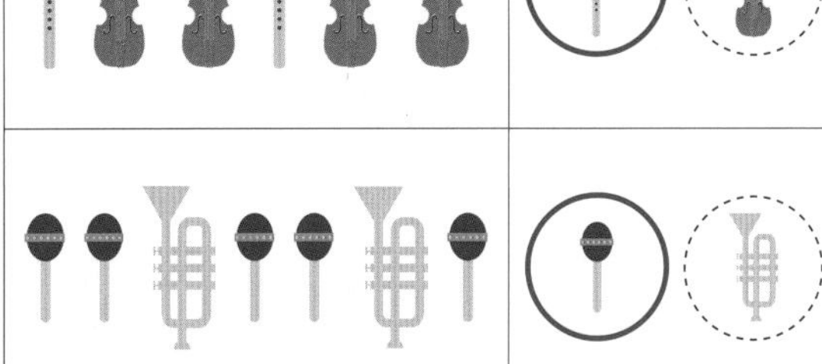

Using this page: Have students look at the instruments in each row, then trace the dashed line around the one that comes next.
Concept: Recognize AB, AAB, and ABB patterns in sound.

34 3-2 Sound Patterns

Exercise 3 • pages 35 – 36

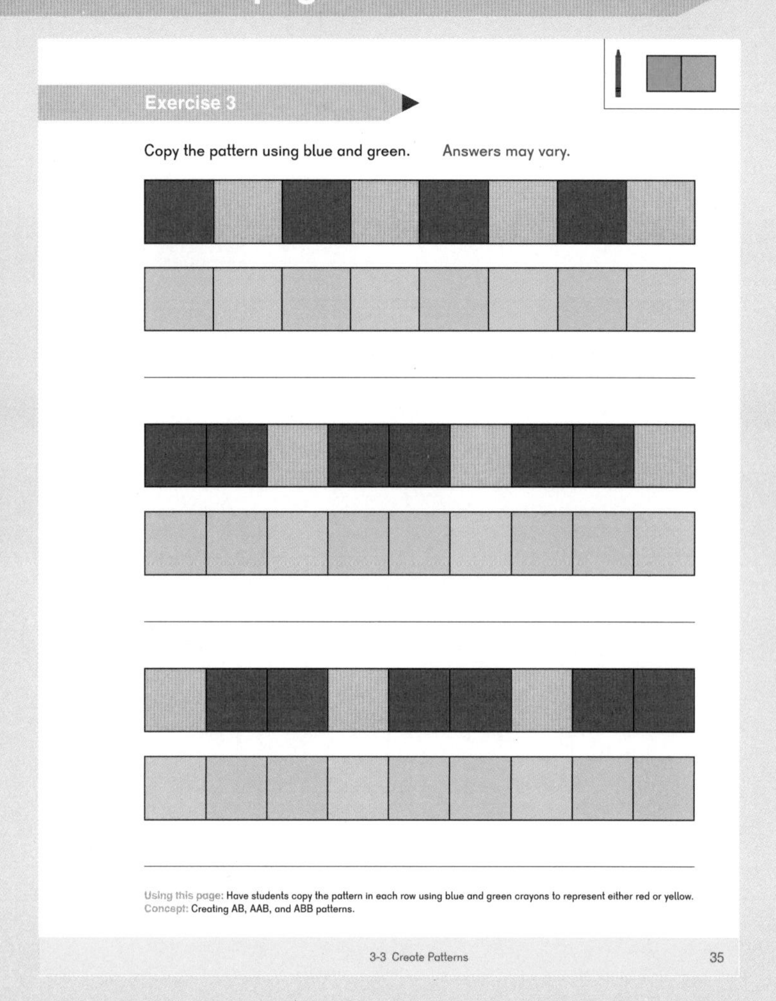
Exercise 3

Copy the pattern using blue and green. Answers may vary.

Using this page: Have students copy the pattern in each row using blue and green crayons to represent either red or yellow.
Concept: Creating AB, AAB, and ABB patterns.

3-3 Create Patterns 35

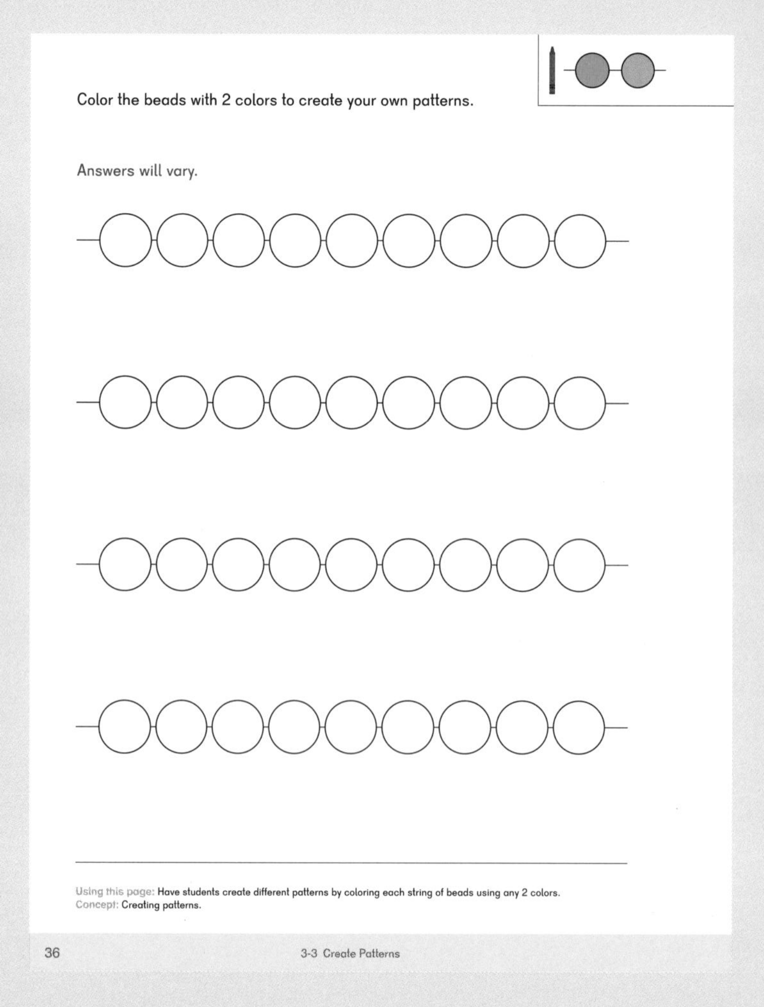
Color the beads with 2 colors to create your own patterns.

Answers will vary.

Using this page: Have students create different patterns by coloring each string of beads using any 2 colors.
Concept: Creating patterns.

36 3-3 Create Patterns

Exercise 4 • pages 37 – 38

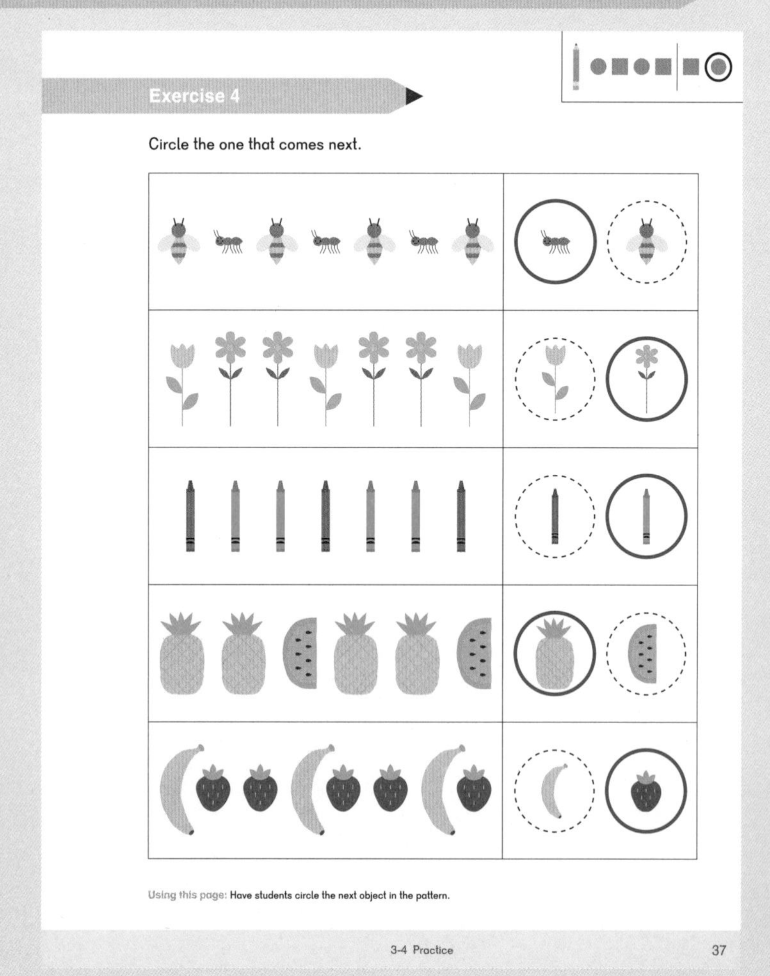
Exercise 4

Circle the one that comes next.

Using this page: Have students circle the next object in the pattern.

3-4 Practice 37

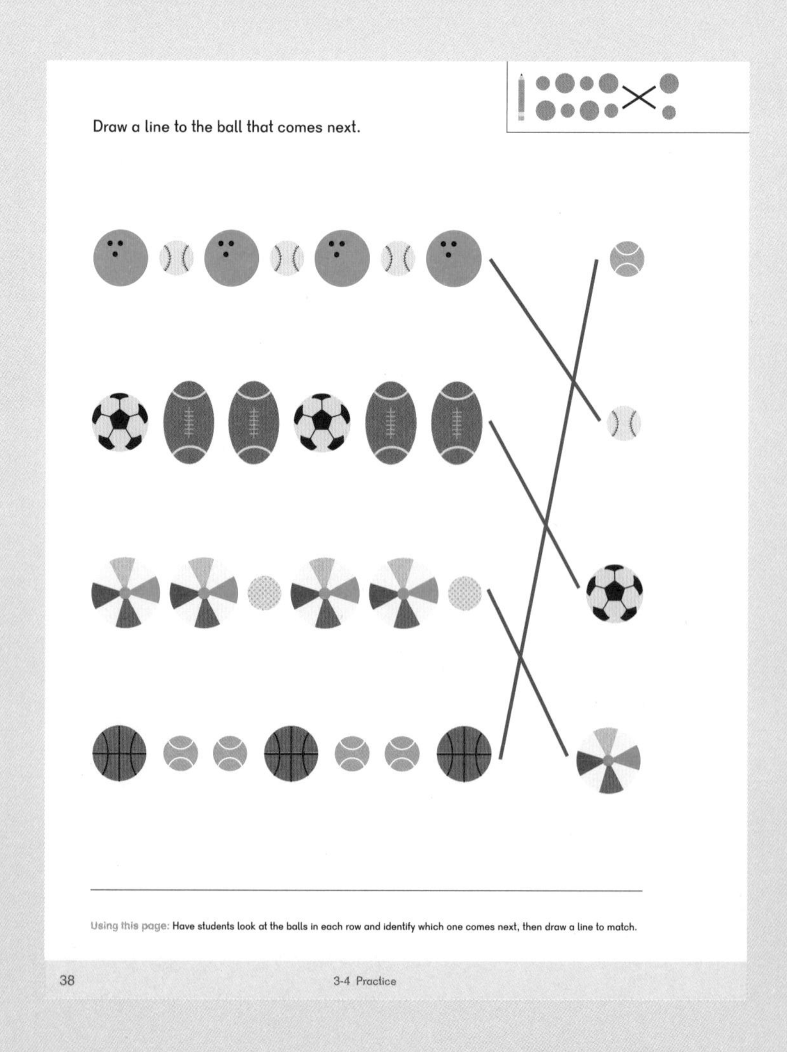
Draw a line to the ball that comes next.

Using this page: Have students look at the balls in each row and identify which one comes next, then draw a line to match.

38 3-4 Practice

Suggested number of class periods: 14 – 15

Lesson		Page	Resources	Objectives
	Chapter Opener	p. 71	TB: p. 45	
1	Count 1 to 5 — Part 1	p. 72	TB: p. 46	Count 1 to 5 by rote.
2	Count 1 to 5 — Part 2	p. 74	TB: p. 48	Count 1 to 5 by rote.
3	Count Back	p. 76	TB: p. 50	Count back from 5 to 1 by rote.
4	Count on and Back	p. 78	TB: p. 52	Review counting from 1 to 5 and from 5 to 1.
5	Count 1 Object	p. 80	TB: p. 54 WB: p. 39	Count 1 object.
6	Count 2 Objects	p. 82	TB: p. 56 WB: p. 41	Count 2 objects with one-to-one correspondence.
7	Count Up to 3 Objects	p. 84	TB: p. 58 WB: p. 43	Count 3 objects with one-to-one correspondence.
8	Count Up to 4 Objects	p. 86	TB: p. 59 WB: p. 45	Count 4 objects with one-to-one correspondence. Recognize the color purple.
9	Count Up to 5 Objects	p. 88	TB: p. 60 WB: p. 49	Count 5 objects with one-to-one correspondence.
10	How Many? — Part 1	p. 90	TB: p. 62 WB: p. 51	Count up to 5 objects with cardinality. Recognize the color pink.
11	How Many? — Part 2	p. 92	TB: p. 64 WB: p. 53	Count up to 5 objects with cardinality.
12	How Many Now? — Part 1	p. 94	TB: p. 66 WB: p. 55	Understand that a rearrangement of objects does not change the number of objects in the set. Recognize the color brown.
13	How Many Now? — Part 2	p. 96	TB: p. 68 WB: p. 57	Understand that a rearrangement of objects does not change the number of objects in the set.
14	Practice	p. 98	TB: p. 70 WB: p. 59	Practice concepts introduced in this chapter.
	Workbook Solutions	p. 101		

Chapter Vocabulary

- One
- Two
- Three
- Four
- Five
- Pair
- Purple
- Five-frame card
- Pink
- Brown
- Rearrangement

Dimensions Math® Pre-K will focus on three types of counting:

- Rote counting – saying the number words in the correct order
- Counting with one-to-one correspondence
- Cardinal counting – understanding that the last number said, when counting a set of objects, is the number of objects in the set

Some students entering Pre-K may already know how to count by rote and can say the numbers in correct order. That skill will be reviewed in this chapter. Proficiency in rote counting will make the more conceptual types of counting learned in this chapter easier.

Counting on, counting up, and counting in ascending order are the same. Counting down, counting back, and counting in descending order are the same. In the **Dimensions Math®** series, the terms "count on" and "count back" are used.

Counting out a specific number of objects from a larger set is more challenging than identifying the number of objects in a set. Counting should be practiced regularly for the remainder of the year.

Key Points

Counting goals for students leaving Pre-K include:

- Rote counting in ascending order from 1 to 10
- Rote counting in descending order from at least 5 to 1
- Counting with one-to-one correspondence up to a set of a minimum of five objects
- Counting with cardinality up to a set of a minimum of five objects

Rote counting lessons include fewer concrete materials due to their verbal nature. Counting with one-to-one correspondence and with cardinality are skills that cannot be taught. Instead, these skills are governed by each child's brain development. Just as children learn to speak by repeated exposure to language, the more exposure and experience a child has with these types of counting, the more likely they are to reach these developmental milestones.

Starting in **Lesson 5: Count 1 Object**, having students name the unit of what they are counting is important. For example, in Lesson 5, when you ask students how many bunnies there are, they should answer, "One bunny." Having students focus on the unit builds on all the matching, grouping, and sorting they did in Chapters 1 and 2. Now students are counting specific objects, for example, bunnies.

New materials which will help students keep counts organized and develop visualization skills are introduced in this chapter.

- five-frame cards
- ten-frame cards

For this reason, students should be required, at least in the beginning, to use their five-frames and ten-frames horizontally, and to start counting from the left. Since we read from left to right and horizontal number lines are written from left to right, this detail is important. As students begin to build understanding and visualization ability, they may be more creative with their use of the cards.

Students will be using fingers to count, starting with their left pinkies. Research indicates that counting on fingers is critical in learning mathematics.[1] Research also indicates that the ability to differentiate between fingers improves numerical performance.[2]

Subitizing, the ability to name a quantity without counting, will be briefly introduced in this chapter. For more information regarding subitizing, see the **Notes** for Chapter 5.

[1]Berteletti, Ilaria; Booth, James R. Perceiving Fingers in Single-digit Arithmetic Problems. Evanston: Cognitive Science, 2015.
[2]Gracia-Bafalluy, M.; Noel, MP. Does Finger Training Increase Young Children's Numerical Performance? Elsevier Masson Srl., 2008.

Where is Thumbkin?

Where is Thumbkin?
Where is Thumbkin?
Here I am. *(Bring right hand to front, with thumb up)*
Here I am. *(Bring left hand to front, with thumb up)*
How are you this morning?
Very well, I thank you. *(Wiggle thumbs as if they're talking to each other)*
Run away. *(Hide right hand behind back)*
Run away. *(Hide left hand behind back)*

Where is Pointer?
Where is Pointer?
Here I am. *(Bring right hand to front, with index finger up)*
Here I am. *(Bring left hand to front, with index finger up)*
How are you this morning?
Very well, I thank you. *(Wiggle fingers as if they're talking to each other)*
Run away. *(Hide right hand behind back)*
Run away. *(Hide left hand behind back)*

Where is Tall Man?
Where is Tall Man?
Here I am. *(Bring right hand to front, with third finger up)*
Here I am. *(Bring left hand to front, with third finger up)*
How are you this morning?
Very well, I thank you. *(Wiggle fingers as if they're talking to each other)*
Run away. *(Hide right hand behind back)*
Run away. *(Hide left hand behind back)*

Where is Ring Man?
Where is Ring Man?
Here I am. *(Bring right hand to front, with fourth finger up)*
Here I am. *(Bring left hand to front, with fourth finger up)*
How are you this morning?
Very well, I thank you. *(Wiggle fingers as if they're talking to each other)*
Run away. *(Hide right hand behind back)*
Run away. *(Hide left hand behind back)*

Where is Pinkie?
Where is Pinkie?
Here I am. *(Bring right hand to front, with Pinkie finger up)*
Here I am. *(Bring left hand to front, with Pinkie finger up)*
How are you this morning?
Very well, I thank you. *(Wiggle fingers as if they're talking to each other)*
Run away. *(Hide right hand behind back)*
Run away. *(Hide left hand behind back)*

Where is the family?
Where is the family?
Here we are. *(Bring right hand to front, with all fingers showing)*
Here we are. *(Bring left hand to front, with all fingers showing)*
How are you this morning?
Very well, we thank you. *(Wiggle fingers as if they're talking to each other)*
Run away. *(Hide right hand behind back)*
Run away. *(Hide left hand behind back)*

Materials

- Yarn/string
- Dot stickers
- Beads
- Cutouts of five-pointed stars
- Video clip of basketball game
- Basketball and basketball hoop
- Raw potatoes and knife
- Music player
- Small animal stickers
- Gloves and mittens
- Socks and toe socks
- Five-divot egg cartons
- Bags
- Keyboard or xylophone (or keyboard app)
- Pitcher
- 5 similar drinking glasses
- Unsharpened pencils
- Dance shoes or tutu
- Swim goggles, mask, or flippers
- Photo of rhesus macaque monkey
- Peas, fresh or frozen
- Linking cubes
- Globe
- Rocks (to be painted)
- Empty containers or pictures of grocery items
- Small and large paper plates
- Objects in different shades of purple, pink, and brown
- Dress-up clothes representing various sports, “grown-up” clothes and baby accessories
- Picture books of animals
- Headphones
- Play fruit or pictures of fruit
- Bowls or baskets
- Brown lunch bags
- Large dried beans
- Small counters
- Index cards
- Tray or box lid
- Work mat

Note: Materials for Activities will be listed in detail in each lesson.

Blackline Masters

- Counting Up to 5 Template
- Five-frame Cards
- Blank Five-frame
- Keyboard Template
- Dot Cards
- Four-column Template

Storybooks

- *The Princess and the Pea* by Hans Christian Andersen
- *The Dot* by Peter H. Reynolds
- *The Three Little Pigs*
- *Hands Down: Counting by Fives* by Michael Dahl

Optional Snacks

- Frozen peas or snap peas
- Honey graham crackers
- Banana slices
- Goldfish® crackers
- Bagels, toppings for bagels
- Dry cereal, including some o-shaped cereal in different colors
- Crackers
- Assorted fruits and vegetables
- Finger sandwiches
- Spaghetti (uncooked; used in preparing snack but not eaten)
- Bean dip and chips
- Purple grapes

Letters Home

- Chapter 4 Letter

Notes

Lesson Materials

- Where is Thumbkin? (VR)
- Gloves and mittens
- Regular socks and toe socks

Note: Pictures of the gloves, mittens, and socks are acceptable.

Explore

Hold up a glove and a mitten and have students say what is the same and what is different about them. Repeat with a regular sock and a toe sock.

Hold up one hand and wiggle your fingers. Have students do the same. Teach them the song, "Where is Thumbkin?" (VR) taught to the tune of "Frère Jacques" (see lyrics and movements on page 68).

Learn

Have students look at page 45 and identify the objects shown. Ask them if they notice anything that is the same about the hand, foot, and star. Then count the fingers, toes, and points on the star with them.

Small Group Center Play

Blocks: Allow students to build as desired.

Dress-up: Include gloves, mittens, socks, and toe socks.

Art: Add pictures of five-pointed stars from Counting Up to 5 Template (BLM) to the art center and allow students to decorate them as desired.

Counting: Provide small objects and five-divot egg cartons. Allow students to explore as desired.

Extend Explore

Basketball: Have students watch a video clip of a basketball game and tell you the number of players playing for each team. Talk to them about some of the rules of basketball. Let them shoot some balls if your school has a basketball hoop.

Materials: Video clip of a basketball game, basketballs, basketball hoop

Lesson 1 Count 1 to 5 — Part 1

Objective

- Count 1 to 5 by rote.

Lesson Materials

- Linking cubes – 1 bag of 5 green cubes for each student
- Keyboard or xylophone, if available
- Count to 5 (VR)
- Optional snack: Snap peas

Explore

Give each student a bag of linking cubes. Tell them that today, they will be pretending that the linking cubes are peas and that the bags are pea pods.

Have students open the bag and take the cubes out, one at a time, slowly, as you do the same. As you remove your "peas," count them.

Teach students the fingerplay "Five Fat Peas" on page 46.

Learn

Have students look at page 46 and tell them that there are five peas in that pod. Have them repeat after you, "One, two, three, four, five peas." Read the rhyme to them again.

Sing (and play, if possible), five notes, low to high, and say, "One – two – three – four – five," with each note. Have students sing with you.

Read aloud Sofia's comments on page 47. Teach students the song "Count to 5" (VR).

Whole Group Play

How Many Ways Can You Count?: Put students into groups of five and have them practice rote counting from 1 to 5 together. Try to put at least one student who already knows how to count to five into each group. They can whisper, shout, count while clapping hands, etc.

I like to count
My own fingers.
I have 5 on each hand.
1 – 2 – 3 – 4 – 5
1 – 2 – 3 – 4 – 5
1 – 2 – 3 – 4 – 5
5 fingers on my hand.
I like to count
My own cute toes.
I have 5 on each foot.
1 – 2 – 3 – 4 – 5
1 – 2 – 3 – 4 – 5
1 – 2 – 3 – 4 – 5
5 cute toes on my foot.

There are 5 of us.
Let's sing about 5.

Objective: Count 1 to 5 by rote.

4-1 Count 1 to 5 — Part 1 47

47

No exercise for this lesson.

Extend Learn

Musical Note Exploration: Have students line the glasses up and fill them with different amounts of water. Then have them arrange the glasses in order of increasing amounts of water from left to right, and play musical notes using the pencil as a mallet. Ask them what they notice about the amounts of water and the pitches of the notes.

Materials: 5 similar glasses, pitcher of water, unsharpened pencil

Small Group Center Play

Blocks: Separate the blocks into groups of five. Have students create structures using five blocks.

Dress-up: Separate the dress-up items into groups of five. Have students use a total of five items for dressing up.

Kitchen: Set out five different types of play food.

Art: Provide paint in the seven colors learned so far: red, blue, yellow, green, orange, black, and white.

Counting: Provide small objects and five-divot egg cartons. Allow students to explore as desired.

Objective

- Count 1 to 5 by rote.

Lesson Materials

- Dance shoes or tutu
- Swim goggles, mask, or flippers
- Optional snack: Honey graham crackers

Explore

Tell the students that Mei likes to do something that requires either dance shoes or a tutu as you hold up each. Have students guess the activity. Repeat for Emma with either goggles, mask, or flippers.

Have students stand and do some dance steps and swim strokes.

Lead students in five dance steps, counting each step as you do it. Repeat with swim strokes.

Have students sit and lead them in clapping five beats, counting each clap. You may choose to change the tempo of the beats as you count and have students follow your lead.

Learn

Have students look at page 48 and discuss Mei's and Emma's comments. Teach students the rhymes.

Have students look at page 49 and discuss the pictures. Teach them the rhymes.

Whole Group Play

Race Cars: Take students outside for a pretend car race. Show them how to be race cars by saying, "Vroom, vroom," and pumping their arms while running. Have them stand in a row on a start line. Designate a finish line. When you say, "On your mark, get set, go!" all students race to the finish.

Count Along: With students in groups of five, have them rote count from 1 to 5 several times out loud.

No exercise for this lesson.

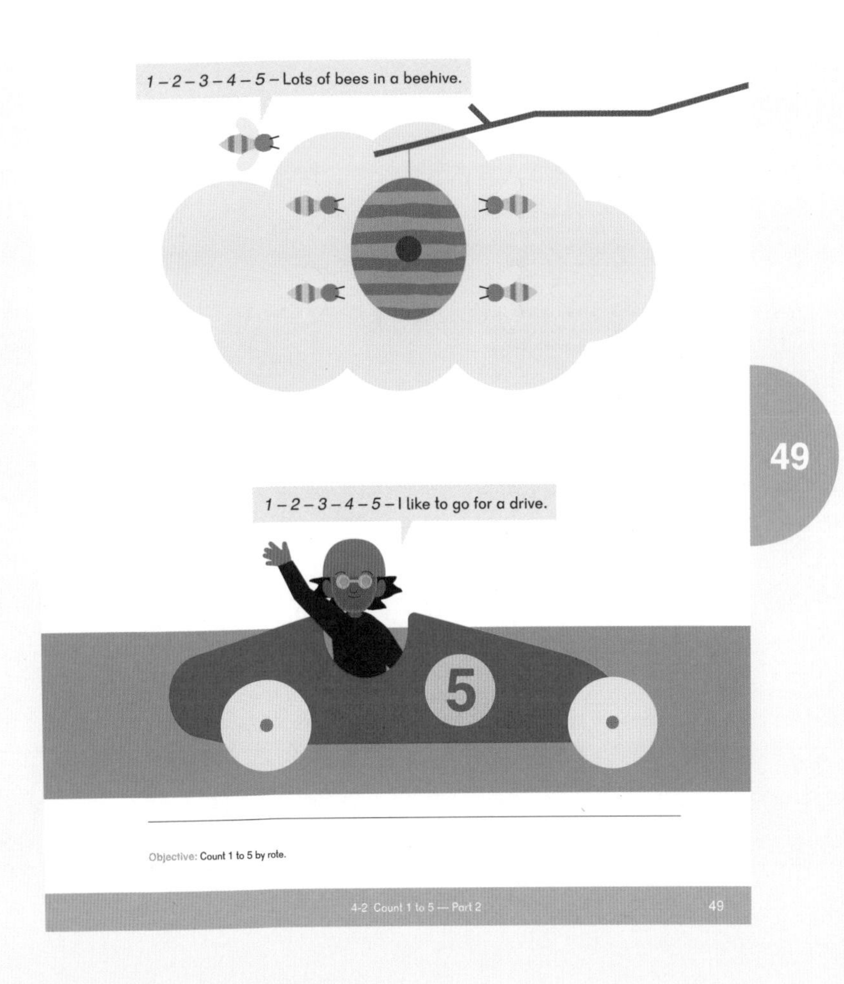

Extend Play

Why We Need Bees: Discuss the importance of bees.

Small Group Center Play

Blocks: Separate the blocks into groups of five. Have students create structures using five blocks.

Busy Bees: Have students pretend to be bees buzzing around a hive.

Art: Provide paint in the seven colors learned so far.

Counting: Provide each student with small objects and a five-divot egg carton. Have them put an object in each divot of one egg carton.

Objective

- Count back from 5 to 1 by rote.

Lesson Materials

- Keyboard or xylophone, if available
- Five Little Monkeys (VR)
- Objects – 5 total
- Optional snack: 5 banana slices

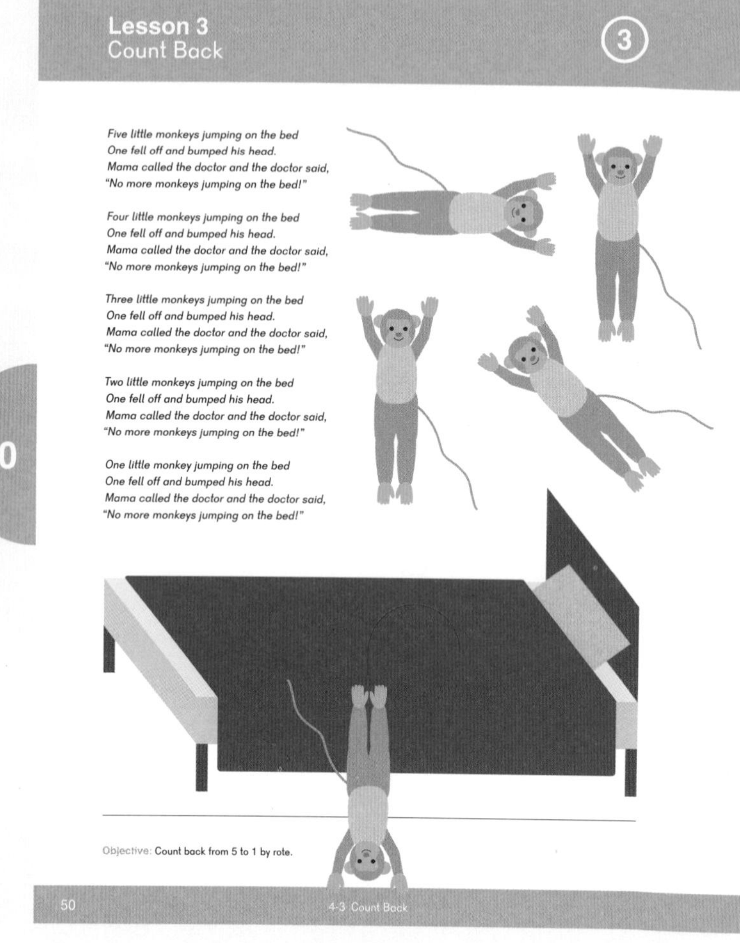

50

Explore

Play and sing five musical notes from low to high as you did in Lesson 1, counting the notes from 1 to 5 as you do so. Have students sing and count with you.

Play and sing five musical notes from high to low, counting from 5 to 1 as you do so. Have students sing and count with you.

Have five students stand up one at a time as you count, "One, two, three, four, five." Have the standing students sit down one at a time as you count backwards, "Five, four, three, two, one." Repeat with other students.

Learn

Have students look at page 50. Read "Five Little Monkeys" to them, or play the video of it (VR). Read the rhyme again, having students act it out.

Have students look at page 51 and discuss the picture. Either watch a video or read "Five Little Ducks" to students. If desired, teach them the song and have them act out the story, doing the duck walk as they do so.

Show students five objects. Have them repeat as you take one object away at a time, "Five, four, three, two, one."

Whole Group Play

Elevator: Have students crouch down as low as possible. As they count from 1 to 5, have them slowly raise up until they are standing tall at five. Then, have them count back from 5 to 1, lowering their bodies as they do so.

Five little ducks
Went out one day
Over the hills and far away.
Mama Duck said,
"Quack, quack, quack, quack."
Only four little ducks came back.

Four little ducks
Went out one day
Over the hills and far away.
Mama Duck said,
"Quack, quack, quack, quack."
Only three little ducks came back.

Three little ducks
Went out one day
Over the hills and far away.
Mama Duck said,
"Quack, quack, quack, quack."
Only two little ducks came back.

Two little ducks
Went out one day
Over the hills and far away.
Mama Duck said,
"Quack, quack, quack, quack."
Only one little duck came back.

One little duck
Went out one day
Over the hills and far away.
Mama Duck said,
"Quack, quack, quack, quack."
No little ducks came waddling back.

No little ducks
Went out one day
Over the hills and far away.
Sad Mama Duck said,
"Quack, quack, quack, quack."
All five ducks came waddling back.

51

Objective: Count back from 5 to 1 by rote.

4-3 Count Back 51

No exercise for this lesson.

Extend Learn

Monkey Count: Tell the students that rhesus macaques have been trained to count and to recognize some numbers. Show them a picture of a rhesus macaque and have them compare the monkeys on page 50 to the picture you show them.

Materials: Picture(s) of rhesus macaques

Small Group Center Play

Blocks: Separate the blocks into groups of five. Have students create structures using five blocks.

Monkey Time: Provide pictures or picture books of monkeys. Have students act like monkeys.

Art: Provide paint in the seven colors learned so far.

123 **Counting:** Provide five-divot egg cartons with a counter in each. Have students remove counters, one at a time, counting back from five.

Objective

- Review counting from 1 to 5 and from 5 to 1.

Lesson Materials

- *The Princess and the Pea* by Hans Christian Andersen
- 5 Peas – real or play
- Magic Thumb (VR)
- Optional snack: Frozen peas or snap peas

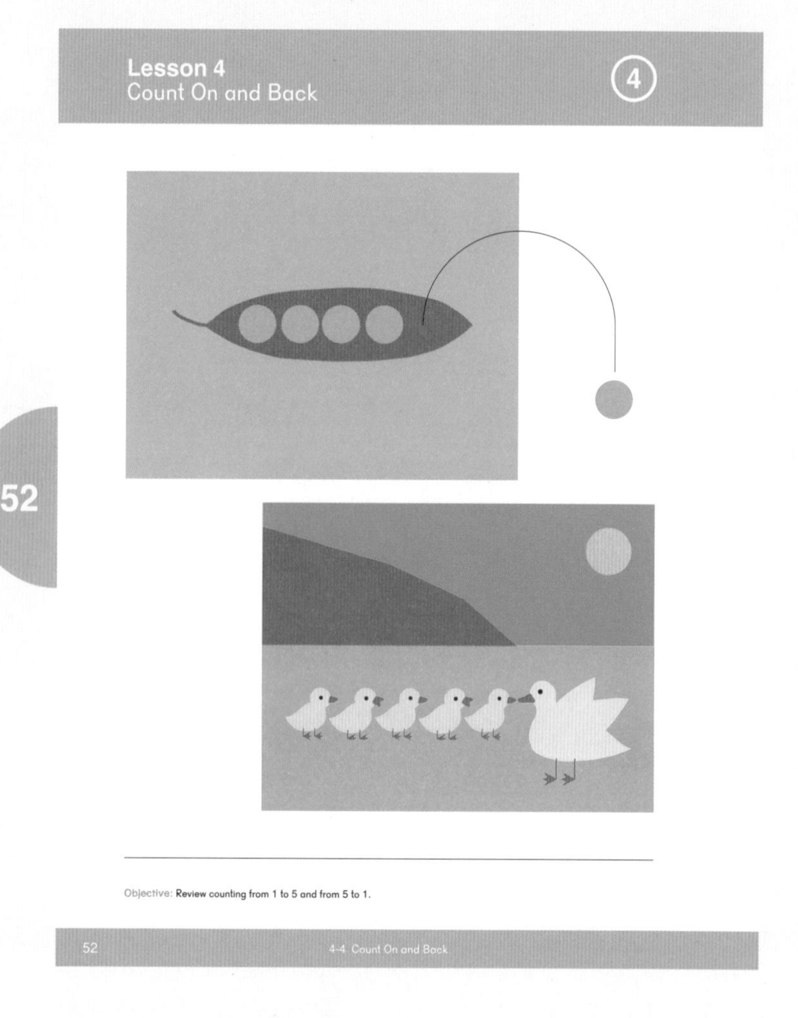

Explore

Read *The Princess and the Pea* aloud. Have students discuss whether sleeping on a mattress on top of a pea would be difficult as you show them one pea. Repeat as you show two, three, four, and five peas.

Read "Five Fat Peas" from textbook page 46 and have students do the finger play with you. Sing "Five Little Ducks" from textbook page 51 and have students act it out.

Show students your thumb and tell them that you are going to teach them a new game called "Magic Thumb" (VR).

Practice counting from 1 to 5, then from 5 to 1, with your thumb pointing in the correct direction. When students become confident, change directions in the middle of the count.

Learn

Have the students look at page 52 and repeat the rhymes with you.

Have students look at page 53. Read Mei's question aloud and have students answer.

Whole Group Play

Run on Your Number: Assign each student a number from 1 to 5 and take them outside. Tell them that as you call a number, the student who has that number in each group must run to a "base" and sit down. Pretend to forget numbers here and there in the sequence, for example, "I already called ones and twos. If I am counting on, which number comes next?"

53

Objective: Review counting from 1 to 5 and from 5 to 1.

4-4 Count On and Back 53

No exercise for this lesson.

Extend Play

Can You Make What I Made?: Give each pair of students 10 blocks. Have each student create a structure using five blocks, one at a time. The other student must build the same structure.

Materials: 10 blocks per pair of students

Small Group Center Play

Blocks: Separate the blocks into groups of five. Create structures using five blocks and have students copy your structure.

Which Friend Are You?: Have students pretend to be Dion, Emma, Alex, Sofia, or Mei. If possible, have dress-up clothes similar in color to those worn by the friends. Students may choose to show activities they have learned that the friend likes to do, such as Emma and swimming.

Art: Have students draw pictures of the five friends.

123 **Sort and Count:** Give students a five-divot egg carton and several types of small counters (at least five of each type). Have students sort the counters by type, then use one type of counter to place a counter in each divot of the egg carton. Have them remove the counters, counting back as they do so, and repeat the activity with a different type of counter.

Objective

- Count 1 object.

Lesson Materials

- Bags containing linking cubes – 1 red, 2 blue, 3 green, 4 white, and 5 black, 1 bag per student
- Optional snack: Goldfish® crackers

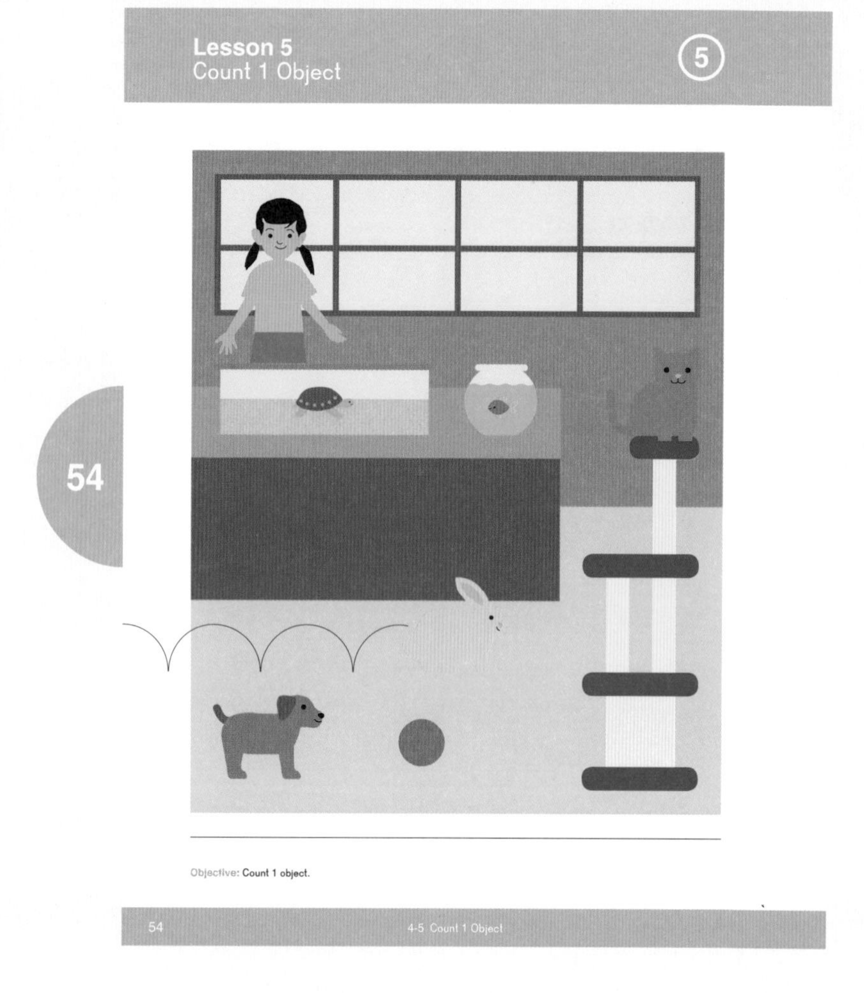

Explore

Give each student a bag containing linking cubes. Have them sort the cubes by color, then say which color cube they have that doesn't match any others.

Point out objects in the classroom of which there is only one. Ask them how many there are. Point to each object and say, "One _____."

Have students look around the classroom from where they are sitting and tell a partner, "I see one _____," while pointing at the object.

Learn

Have students look at page 54 and ask them how many of each animal they see. Have them touch each picture as they answer and say, "I see one _____."

Have students look at page 55 and say how many circles there are. Read Sofia's direction aloud and have them color the circle.

Whole Group Play

I Found 1: Send students on a hunt to bring one object back to class. If weather allows, take students outside, otherwise have them hunt in the classroom. Then have each student "show and tell."

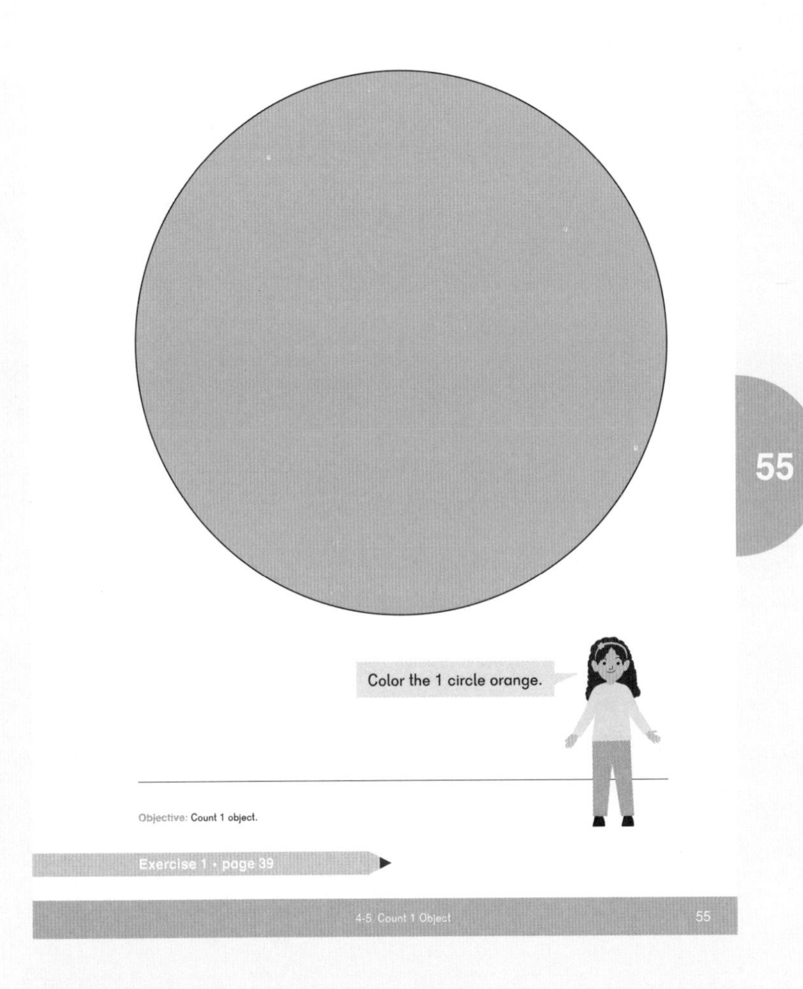

Small Group Center Play

Blocks: Provide different types of blocks, one of each type. Examples: one DUPLO® or LEGO® type block of each size and color available; one wooden block of each size and type available. Allow students to explore as they wish.

Dress-up: Include only one of each type of clothing, object, etc.

Kitchen: Include only one of each object.

Art: Have students draw a picture of one object using one color.

Sort and Count: Provide a container holding several different types of small counters, several of each type. Have students sort the counters by type, then show one of each type of counter on a work mat.

Exercise 1 • page 39

Extend Learn

Uno, Un, Eins: Have students look at a globe and help them find Italy, Spain, France, and Germany. Tell them that these countries are on the continent of Europe. Tell them how the word "one" is said in some different languages in Europe and write each word for students to see.

Italian and Spanish – uno
French – un
German – eins

Materials: Globe

The Dot: Read *The Dot*. Celebrate International Dot Day (usually held on September 15) by having students draw one dot and turn it into art.

Materials: *The Dot* by Peter H. Reynolds

Goldfish® is a trademark of Pepperidge Farm.
DUPLO® and LEGO® are trademarks of the LEGO Group of companies.

Objective

- Count 2 objects with one-to-one correspondence.

Lesson Materials

- Bags containing linking cubes – 2 red, 3 blue, 4 yellow, and 5 black (colors for example only), 1 bag per pair of students
- Objects – matching pairs, several different types, e.g. 2 toy cars and 2 spoons
- Crayons
- Optional snack: Bagel halves with 2 toppings

56

Lesson 6
Count 2 Objects

Objective: Count 2 objects with one-to-one correspondence.

56 4-6 Count 2 Objects

Explore

Give each pair of students a bag of linking cubes. Have them sort by color. Teach students the word “pair” and tell them that pairs are two objects that match. Demonstrate by holding up two matching objects and counting, “One, two.” Repeat and have students count with you. Ask them to show pairs of linking cubes.

Point out objects in the classroom of which there are only two. Ask them how many there are. Point to each object and say, “One, two _____.”

Have students look around the classroom from where they are sitting and tell a partner, “I see two _____,” while pointing at the objects.

Learn

Have students look at page 56 and identify the objects. Read Alex’s question aloud and have them answer.

Have students look at page 57. Read Sofia’s direction aloud. Draw a circle on the board. Use your finger to draw a circle in the air and have students do the same. Have them complete the task.

Whole Group Play

Pair Game: Select a partner and model the game for the students. Make a funny face. Tell your partner to make the same funny face. Now tell your partner to make a funny face which you mirror.

Have all students stand and find a partner. Partners face each other and play by taking turns making faces and copying.

After playing for a few minutes, have students hold hands with their partners. Tell them that those students holding hands are a pair of students and ask, “How many students are in each pair?” If there is a student without a partner, be that student’s partner.

Draw 2 big circles and color each of them 2 different colors.

Answers will vary.

Objective: Count 2 objects with one-to-one correspondence.

Exercise 2 • page 41

4-6 Count 2 Objects 57

Small Group Center Play

Blocks: Provide different types of blocks, two of each type. Allow students to explore as they wish.

Kitchen: Include only two of each object. If possible, project page 56. Include items similar to those on the page. Have students find objects similar to those shown.

Art: Have students paint a picture of two objects using two colors.

Sort and Count: Provide a container holding several different types of small counters, several of each type. Have students sort the counters by type, then show two of each type of counter on a work mat.

Exercise 2 • page 41

Extend Play

Name that Pair: Have students name as many things as they can that come in pairs.

Lesson 7 Count Up to 3 Objects

Objective

- Count 3 objects with one-to-one correspondence.

Lesson Materials

- Familiar grocery items (or pictures) – several different types, up to 3 of each type
- Bags containing 1 red, 2 black, and 3 blue linking cubes (colors are given for example), 1 bag per student
- Optional snack: Cereal or 3 crackers

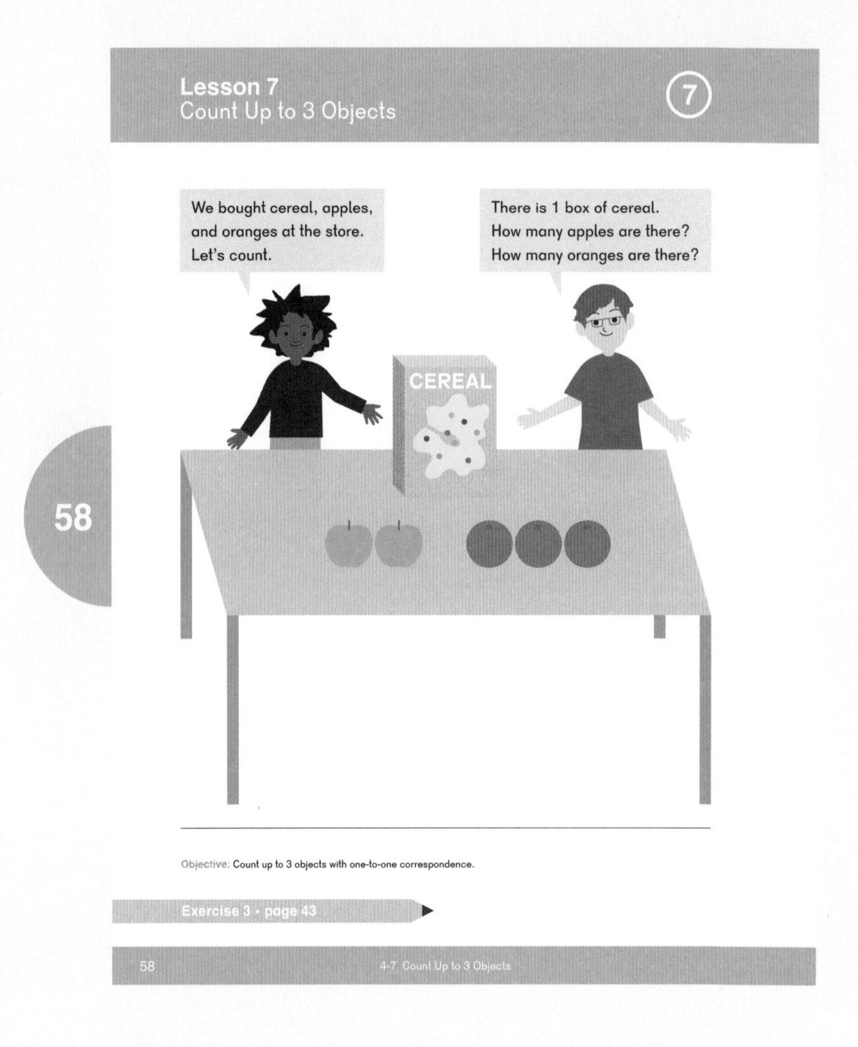

Explore

Show students familiar grocery items or pictures of them. Show one of one type, two of a different type, and three of a third type. Have students identify the objects and discuss their experiences with grocery shopping.

Provide bags of linking cubes to students. Ask them to find the red cube and say, "Count how many red linking cubes you have. Put all of them in front of you." Repeat for black and blue cubes.

Think aloud while counting three blue linking cubes, showing each cube as it is counted. For example, "One – two – three. I will stop counting at three because that's all of the blue linking cubes I have." Have students count their blue linking cubes as you did. Ask students to make towers of the black linking cubes and blue linking cubes, place them next to the red linking cube, and have a brief discussion of which tower is the tallest and which is the shortest.

Learn

Read Dion's statement, then Emma's statement and questions. Ask students how many of each grocery item they see.

Note: Some students may be able to answer without counting.

Whole Group Play

Count and Move: Have students stand in a circle and lead them in an activity where they will do counting motions. For example, "Let's touch our toes three times, 1, 2, 3. Let's tap our head two times, 1, 2. Let's clap our hands one time, 1." Other motions could include hopping, jumping, kneeling, and sitting.

Exercise 3 • page 43

Extend Explore

Reading Time: Read *The Three Little Pigs*. Discuss the personality traits of the pigs and how that affected the outcome of the story.

Materials: *The Three Little Pigs*

Small Group Center Play

Blocks: Provide different types of blocks, three of each type. Allow students to explore as they wish.

Grocery Store: Use play food, up to three of each type, and have students play store.

Potato Stamping: Cut potatoes in half. Use a knife to make two deep grooves in each potato half. Use three different colors of paint, one for each section. Transfer the paint to paper. Show students what you did and leave your picture at the center as a model. Have students create three-color patterns using colors of their choice.

Sort and Count: Provide a container holding several different types of small counters, several of each type. Have students sort the counters by type, then show three of each type of counter on a work mat.

Objectives

- Count 4 objects with one-to-one correspondence.
- Recognize the color purple.

Lesson Materials

- Bags containing linking cubes — At least 4 each of 4 different colors, 1 bag per student
- Four-column Template (BLM)
- Objects — several shades of purple
- Crayons — purple, yellow, and green
- Optional snack: 4 small pieces of fruit

Explore

Provide each student with a bag of linking cubes and a Four-column Template (BLM) . Have students use their templates to put one linking cube in the first column, two of a different color in the second, etc.

Ask students to count how many linking cubes are in the first column. Have them place a finger on the linking cube and say, "One linking cube." Repeat for the second and third column. For the fourth column, count, "One, two, three, four linking cubes," with students.

Introduce the color purple by showing students examples of purple objects. Have them point to other objects in the classroom that are purple. Count a set of four purple objects together.

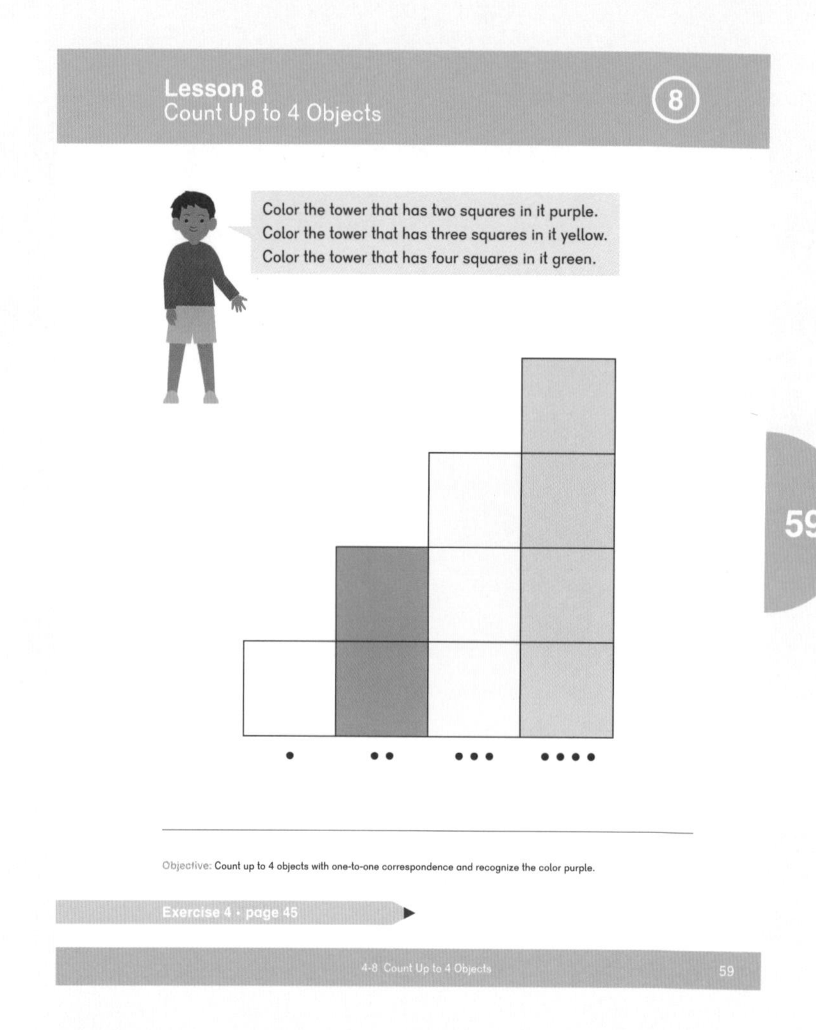

Learn

Discuss page 59. Ask students to count the number of blocks in each column. Ask them to tell you why the dots are at the bottom of each column. Read Alex's directions to them and have them complete the task.

Whole Group Play

Freeze Four: Put students into groups of four. Play music and have students dance. When the music stops, all students must freeze in their groups. Ask them, "How many students are in each group? How do you know?" Count the students in each group.

Materials: Music

Small Group Center Play

Blocks: Have students build towers, each with four blocks.

Grocery Store: Use play food, up to four of each type, and have students play store.

Art: Provide four different paint colors, including purple.

Sort and Count: Provide a container holding several different types of small counters, several of each type. Have students sort the counters by type, then show four of each type of counter on a work mat.

Exercise 4 • page 45

Extend Learn

2 Legs or 4?: Have students look at picture books and discuss animals that walk on four legs and animals that walk on two legs. Have them compare the different modes of movement.

Materials: Picture books about animals

Objective

- Count 5 objects with one-to-one correspondence.

Lesson Materials

- Cutout of five-pointed stars from Counting Up to 5 Template (BLM), 1 per student
- Five-frame Cards (BLM) 1 to 5
- Blank Five-frames (BLM), 1 per student
- Small counters, at least 5 per student
- Crayons
- Optional snack: 5 star-shaped crackers

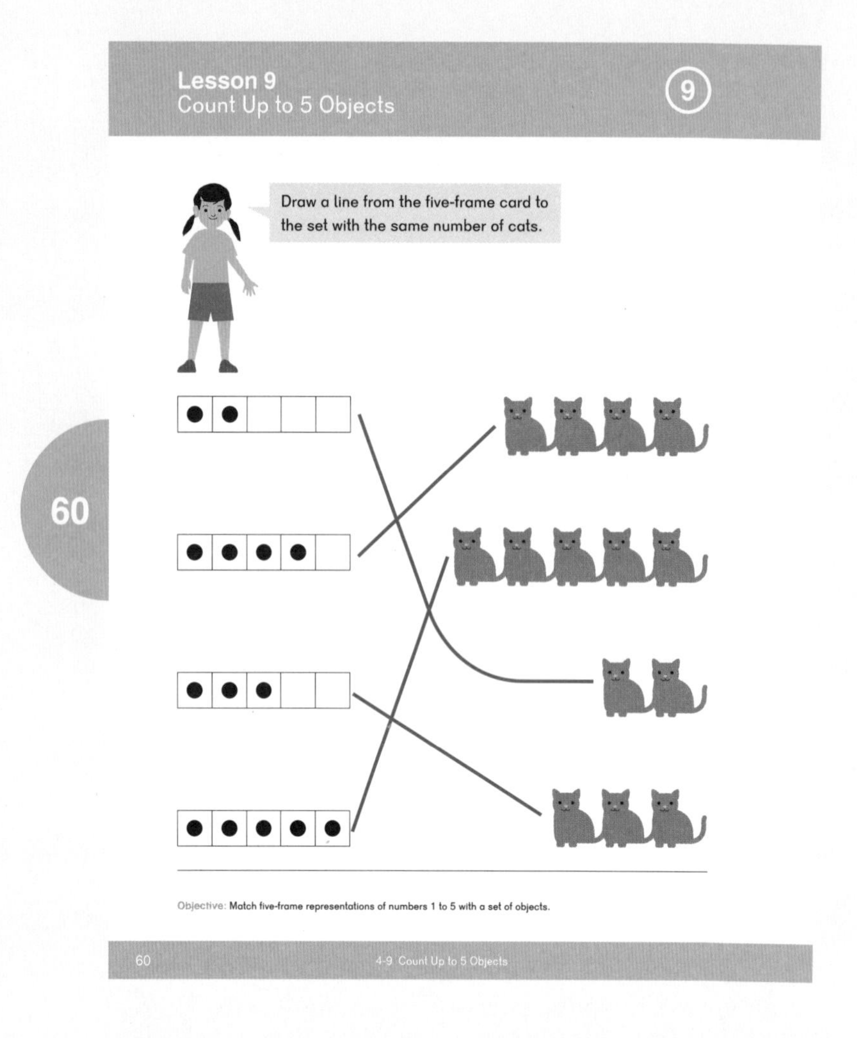

Explore

Tell the students this story as you hold up one of the star cutouts: "Some children were counting the points on a star. Some of them said that there were three points. Others said that there were four points, while others said there were five points."

Provide each pair of students a star and have them discuss a good way to count the points on the star with their partner. Bring attention back to the whole group and discuss strategies for counting. Assess cardinal counting by noticing how many students re-count to say that there are five points on the star.

Introduce Five-frame Cards (BLM) to students with a card that has all five dots colored in. Hold the card horizontally and count the colored dots, having students count them with you. Repeat with different cards showing different numbers of colored dots from 1 to 5.

Give each student a Counting Up to 5 Template (BLM), a Blank Five-Frame (BLM), and some counters. Have them put a counter on each point of their cutouts, then move the counters to their five-frames, counting each point as they move each counter. Collect the counters and star templates.

Learn

Have students look at page 60. Have them say what number is shown on the first five-frame card. Have them use a finger to match the five-frame card to the two cats. Read Mei's directions aloud and have them complete the task.

Have students look at page 61 and identify the objects. Tell them that they have learned more about the friends (Mei likes cats and Alex likes t-ball). Read Alex's directions aloud and have students complete the task.

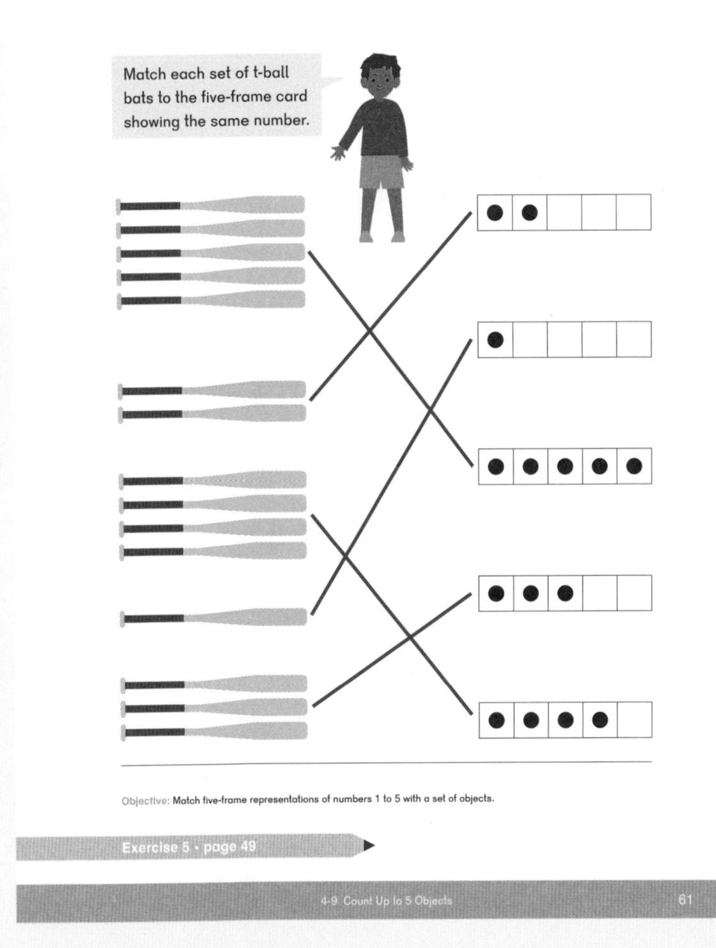

61

Whole Group Play

Five-frame Fun Cover and Switch: Give each student a Five-frame Card (BLM) 1 to 5. Each student's task will be to cover the colored dots with counters, counting each counter as it is placed on the card. Check cards. Have students remove the counters. Play music and have students pass their cards to the left around the circle until the music stops. Students repeat the counting activity with the new cards.

Materials: Five-frame Cards (BLM) 1 to 5 – one card per student, 5 counters for each student, music

How Many?: Five-frame Cards (BLM) with different numbers of dots filled in. Have students call out the number of filled-in dots on each card.

Materials: Five-frame Cards (BLM) 1 to 5

Small Group Center Play

Blocks: Have students build towers, each containing five blocks.

Pets Alive: Use stuffed toys, up to five of each type, and have students talk for the animals.

Rock Art: Provide five different colors of paint, and five different sizes of rocks. Have students dip the rocks in the paint and transfer it to paper.

Sort and Count: Provide a container holding several different types of small counters, several of each type. Have students sort the counters by type, then show five of each type of counter on a work mat.

Exercise 5 • page 49

Extend Learn

Rock Art Patterns: Have students create patterns using paint colors transferred from rocks.

Materials: Art paper, paint, rocks

Reading Time: Read *Hands Down: Counting by Fives.*

Materials: *Hands Down: Counting by Fives* by Michael Dahl

Objectives

- Count up to 5 objects with cardinality.
- Recognize the color pink.

Lesson Materials

- Bags containing 6 cubes or counters, 1 bag per student
- Blank Five-frame (BLM) – 1 per student
- Keyboards, keyboard apps on handheld devices, or Keyboard Templates (BLM)
- Objects – several shades of pink
- Optional snack: Finger Sandwiches

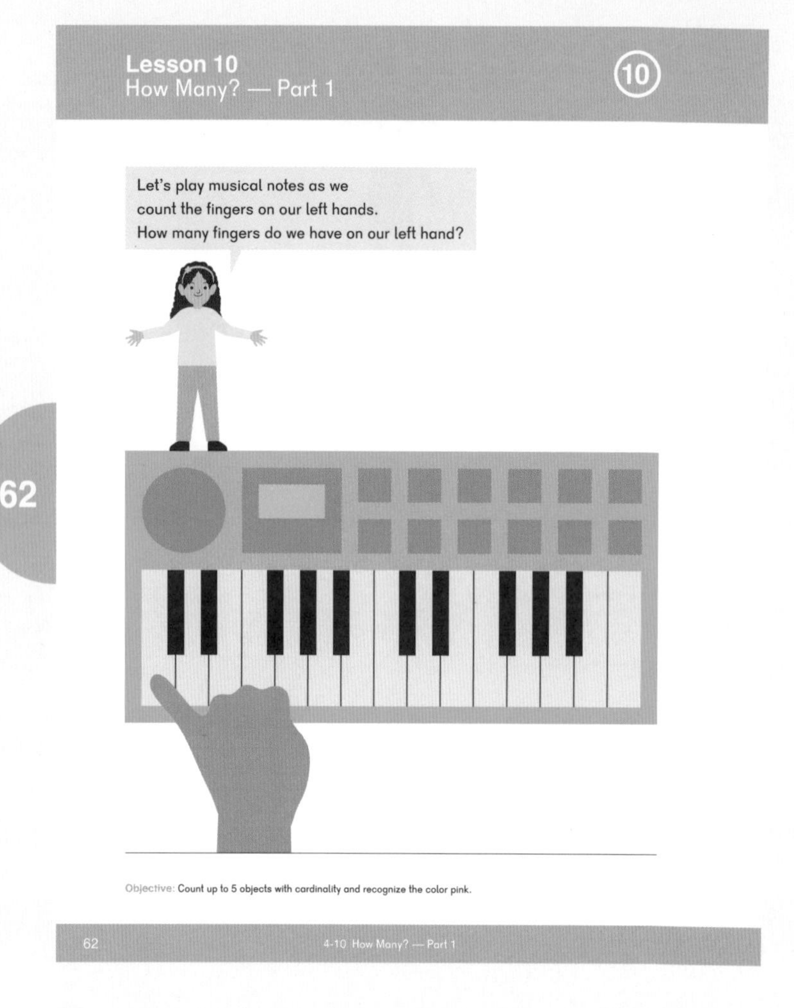

Explore

Provide each student with a bag of counters and a Blank Five-frame (BLM). Have students place one counter on their five-frame. Be sure they place their five-frame horizontally and that they put the counter on the left side of the five-frame. Ask students how many counters they are showing. Have them remove that counter and show two. Repeat to five.

Collect the counters and five-frames and give each student a Keyboard Template (BLM). Draw a keyboard, or project one, and shade in a C. Help students place their left pinkies on C.

Have them play C with their left pinkies, then play D, E, F, and G, respectively, with their left ring, middle, index fingers, and thumb. Repeat, counting each note as it is played. Play or sing number words with each note. After playing the last note, ask, "What number was the last number we said when we counted our notes? How many notes did we play?" Repeat.

Finger Counting (VR): Teach students to count their fingers from 1 to 5 on their left hand.
To do this, stand in front of them and hold up your right hand, palm facing them. Tell them to help you count the fingers on your hand. Hold up your pinkie and say, "One." Continue counting by holding up your ring finger, middle finger, etc, saying, "Two, three, ..."

Have students hold up their left hands, palm facing out. They start counting from their left pinkie. At the end, ask, "How many fingers do we have on our left hand?"

Introduce the color pink by showing students pink objects.

Let's count the toes on our left foot starting with our little toe.
How many toes do we have on our left foot?

63

Objective: Count up to 5 objects with cardinality.

Exercise 6 • page 51

4-10 How Many? — Part 1 63

Learn

Have students look at page 62 and say the colors of the keyboard. Read Sofia's comment and question aloud and have students answer.

Have students look at page 63 and say what Dion was doing before he took his shoe and sock off. Tell them that they have just learned one of Dion's favorite things to do. Read Dion's comment and question and have students answer the question.

Whole Group Play

Finger Counting: Call out numbers 1 to 5 and show students that many fingers, starting with your right pinkie. Students mirror by starting with their left pinkies.

Small Group Center Play

Blocks: Have students count out five blocks and use them to create structures.

Dress-up: Include clothes used in playing various sports.

Yarn Art: Provide pieces of yarn of five different lengths and paint of five different colors, including pink. Have students use yarn to make five paint lines on paper.

123 **Sort and Count:** Provide a container holding several different types of small counters, several of each type. Have students sort the counters by type, then show four, then five of each type of counter on a work mat.

Exercise 6 • page 51

Extend Explore

Keyboard Exploration: Allow students to explore the sounds they make on a keyboard. Ask them what happens to the sounds as they move from left to right, then from right to left on the keyboard.

Materials: Keyboards, headphones

Objective

- Count up to 5 objects with cardinality.

Lesson Materials

- Fruit – real or play pieces, at least 5 of each type, of colors learned so far
- Containers – enough for each unique type of fruit
- Optional snack: Fresh fruit or vegetables

Lesson 11
How Many? — Part 2

64

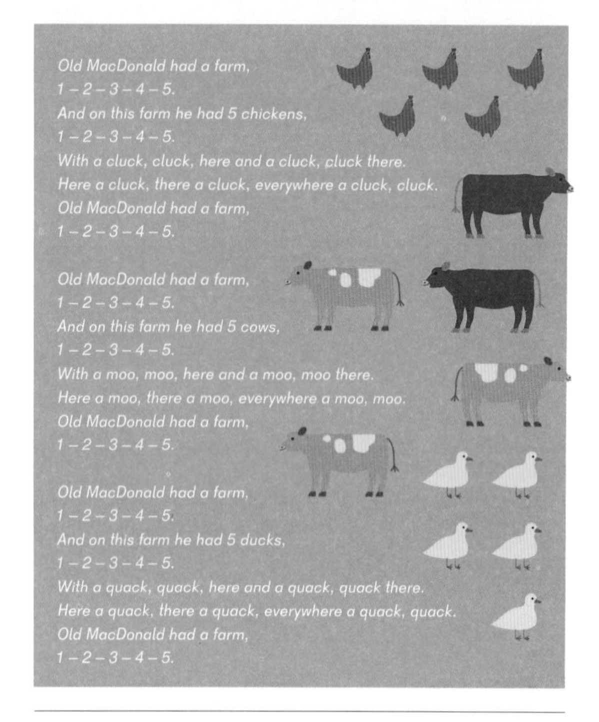

Old MacDonald had a farm,
1 – 2 – 3 – 4 – 5.
And on this farm he had 5 chickens,
1 – 2 – 3 – 4 – 5.
With a cluck, cluck, here and a cluck, cluck there.
Here a cluck, there a cluck, everywhere a cluck, cluck.
Old MacDonald had a farm,
1 – 2 – 3 – 4 – 5.

Old MacDonald had a farm,
1 – 2 – 3 – 4 – 5.
And on this farm he had 5 cows,
1 – 2 – 3 – 4 – 5.
With a moo, moo, here and a moo, moo there.
Here a moo, there a moo, everywhere a moo, moo.
Old MacDonald had a farm,
1 – 2 – 3 – 4 – 5.

Old MacDonald had a farm,
1 – 2 – 3 – 4 – 5.
And on this farm he had 5 ducks,
1 – 2 – 3 – 4 – 5.
With a quack, quack, here and a quack, quack there.
Here a quack, there a quack, everywhere a quack, quack.
Old MacDonald had a farm,
1 – 2 – 3 – 4 – 5.

Objective: Count up to 5 objects with cardinality.

64 4-11 How Many? — Part 2

Explore

Teach students the modified version of "Old MacDonald Had a Farm" as shown on page 64.

Tell students that farmers grow crops in addition to raising animals. Hold up each of the types of fruit you have and ask students to identify the fruit. Ask them to describe experiences they have had shopping for fruits and vegetables.

Place each type of fruit, or pictures of types of fruit, in a separate container. Tell students that they will have an opportunity to "buy" fruit by telling you how many pieces of fruit they would like, 1 to 5, and counting out that many. Model this activity for them.

Learn

Have students look at page 64 and count each type of animal. Read the words aloud to the students, having them use their fingers under the words as you read. This will be an introduction to recognizing numerals.

Have students look at page 65 and identify the types of fruit. Read Alex's comment and direction, and have students count and say how many pieces of each type of fruit there are.

Whole Group Play

Animal Count: Allow students to create works of art using no more than five of any one type of animal. Then have them tell you, for example, "I used two cows and three horses."

Materials: Art paper, animal stickers

Finger Counting: Call out numbers 1 to 5 and show students that many fingers, starting with your right pinkie. Students mirror by starting with their left pinkies.

65

Exercise 7 • page 53

Extend Play

Fruits & Veggies: Have students discuss experiences in which they have gotten fruits and vegetables other than buying them at the grocery store. Do any of their families have gardens? Have they been to a farm?

Small Group Center Play

Blocks: Have students build towers, each containing five blocks.

Farm Store: Give each student a brown lunch bag. Place each type of real or play pieces of fruit (or picture cards of fruit), at least five of each type, in a separate bowl. Have students tell you how many pieces of fruit they would like, up to five. Students must count out the correct number.

Still Life: Set up a bowl containing real or play fruits of the colors learned so far. Have students use crayons or paint to illustrate the arrangement.

Sort and Count: Provide a container holding several different types of small counters, several of each type. Have students sort the counters by type, then show four, then five, of each type of counter on a work mat.

Objectives

- Understand that rearrangement of objects does not change the number of objects in the set.
- Recognize the color brown.

Lesson Materials

- Objects – several shades of brown
- Bag containing 6 beads, 1 per student
- Large paper plates, 1 per student
- String, 1 piece per student, with a knot tied at one end
- Optional Snack: Breakfast cereal with holes, different colors, and uncooked spaghetti (let students make a pattern stick and the eat the cereal)

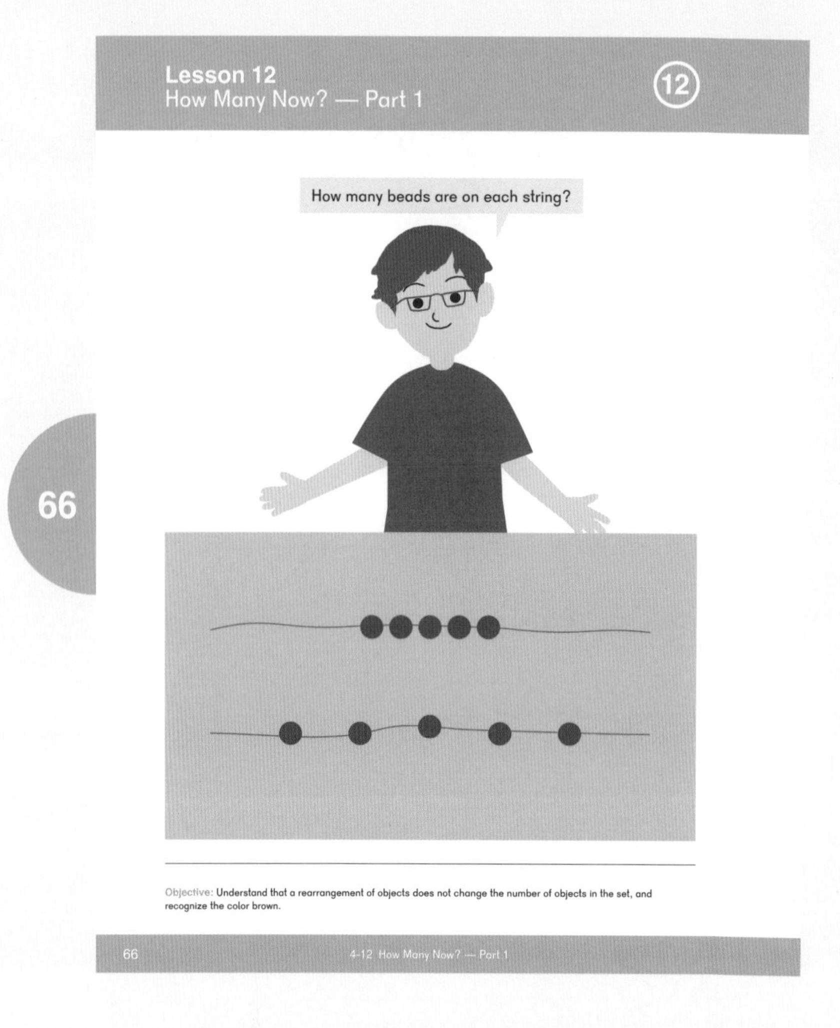

Explore

Introduce the color brown by showing brown objects of different shades. Have students identify other brown objects in the classroom.

Give each student a bag of beads, a paper plate, and a piece of string with a knot tied at one end. Have them count out five beads and put them on the string as you do the same.

Count the beads again, once on the string, then turn the string over so the beads fall off in front of you on to a paper plate. Ask students how many beads are on the paper plate. Count them together, moving each bead as it is counted. Ask students who knew there were five beads on the plate without counting to explain how they knew.

Have students work in pairs. One student in each pair will count the number of beads you call out (three, four, or five) and put them on the string. Both students then count the beads on the string. Next, one of them turns the string so that the beads fall off the string onto the paper plate. The other student will then say the number of beads on the paper plate. Students take turns stringing the beads and pouring them on the plate.

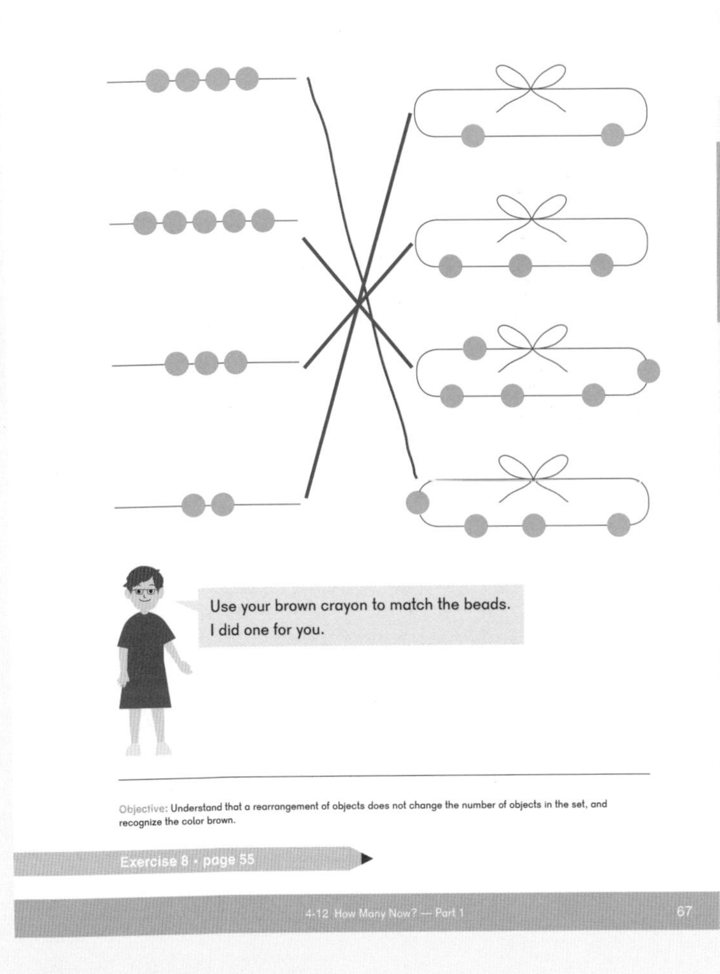

67

Small Group Center Play

Blocks: Place three, four, or five blocks in brown lunch bags. Have students work in pairs. One student pours the blocks out. The other student identifies the number of blocks.

Note: Some students will be able to tell the number without counting.

Bead Store: Same rules as **Farm Store** except students "buy" beads.

Still Life: Set up a bowl containing three, four, or five pieces of real or play fruits of the colors learned so far. Have students use crayons or paint to illustrate the arrangement.

Sort and Count: Provide a container holding several different types of small counters, several of each type. Have students sort the counters by type, then show five of each type of counter on a work mat.

Exercise 8 • page 55

Whole Group Play

Five-frame Flash: With students sitting in a circle, flash Five-frame Cards (BLM) showing different numbers 1 to 5 and have them call out the number shown on each card.

Materials: Five-frame Cards (BLM) 1 to 5

Finger Counting: Call out a number 1 to 5 and have students show you the correct number of fingers starting with their left pinkies.

Extend Learn

Dot Card Match: Set up stations with cards set out faceup. Send small groups of students to each station. Students will take turns matching a Five-frame Card (BLM) to a Dot Card (BLM) showing the same number.

Materials: 2 sets of Five-frame Cards (BLM) 2 to 5, 2 sets of Dot Cards (BLM) 2 to 5

Lesson 13 How Many Now? — Part 2

Objective

- Understand that rearrangement of objects does not change the number of objects in the set.

Lesson Materials

- Small bag
- 5 large dried beans
- Large paper plates, 1 per student
- Black crayon
- Optional snack: Bean dip and chips

Lesson 13
How Many Now? — Part 2

68

Objective: Understand that a rearrangement of objects does not change the number of objects in the set.

68 4-13 How Many Now? — Part 2

Explore

Play "Spill the Beans." Sitting in a circle, have students help you count four beans as you place them in a small bag. Say, "Spill the beans," and spill out the beans onto a paper plate.

Ask students how many beans are on the plate. Have them help you count the beans as you pick them up and place them back in the bag.

Give each student a paper plate. Pass the bag containing the beans to the student to your left. Have students continue passing the bag until you say, *"Spill the beans."* At that time, whichever student is holding the bag spills the beans onto a paper plate and all students say, without counting, how many beans are on the plate. Have that student replace the beans and bring the bag to you.

Repeat with different numbers of beans up to five.

Learn

Have students look at page 68 and say what happened to Dion's tower. Read Dion's questions. Have students place a finger on each block as they first count the blocks in the tower and then the scattered blocks.

Have students look at page 69. Read Dion's direction aloud and have them complete the task.

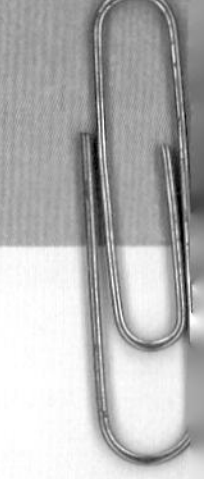

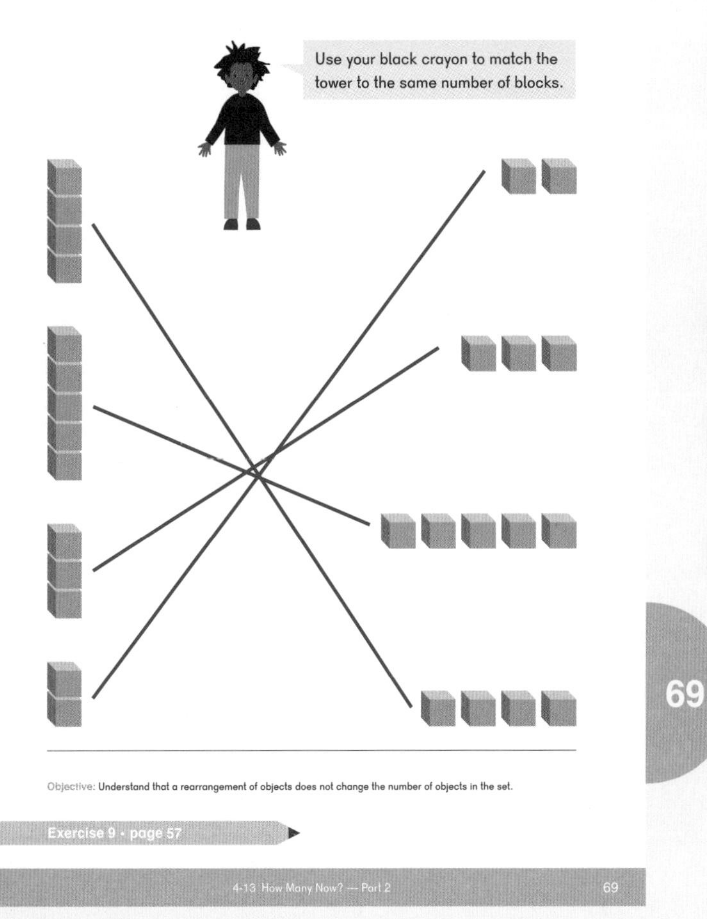

69

Small Group Center Play

Blocks: With students working in pairs, have them take turns where one builds a tower of three, four, or five blocks and the other student tells how many blocks are used. Students then count together to verify.

Dress-up: Include clothes that babies and children would wear.

Art: Make towers of two, three, four, and five blocks. Have students draw each tower.

Sort and Count: Provide a container holding several different types of small counters, several of each type. Have students sort the counters by type, then show four or five of each type of counter on a work mat.

Exercise 9 • page 57

Whole Group Play

Five-frame Flash: With students sitting in a circle, flash Five-frame Cards (BLM) showing different numbers 1 to 5 and have them call out the number shown on each card. Then have them show you the correct number of fingers.

Materials: Five-frame Cards (BLM) 1 to 5

Dot Card Challenge: Have students make their own dot cards. Challenge them to show three dots in at least three different ways, four dots in at least four different ways, and/or five dots in at least five different ways.

Materials: Index cards, dot stickers

Extend Learn

What Did You See?: Play a visual memory game with the students. Put 3 objects on a tray and let students look at the objects for a few seconds, then cover them up and have students tell you what objects they remember. Increase the number of objects as their skills increase.

Materials: Tray or box lid, unique objects, dishtowel to cover tray

Objective

- Practice concepts introduced in this chapter.

Lesson Materials

- Optional snack: Purple grapes

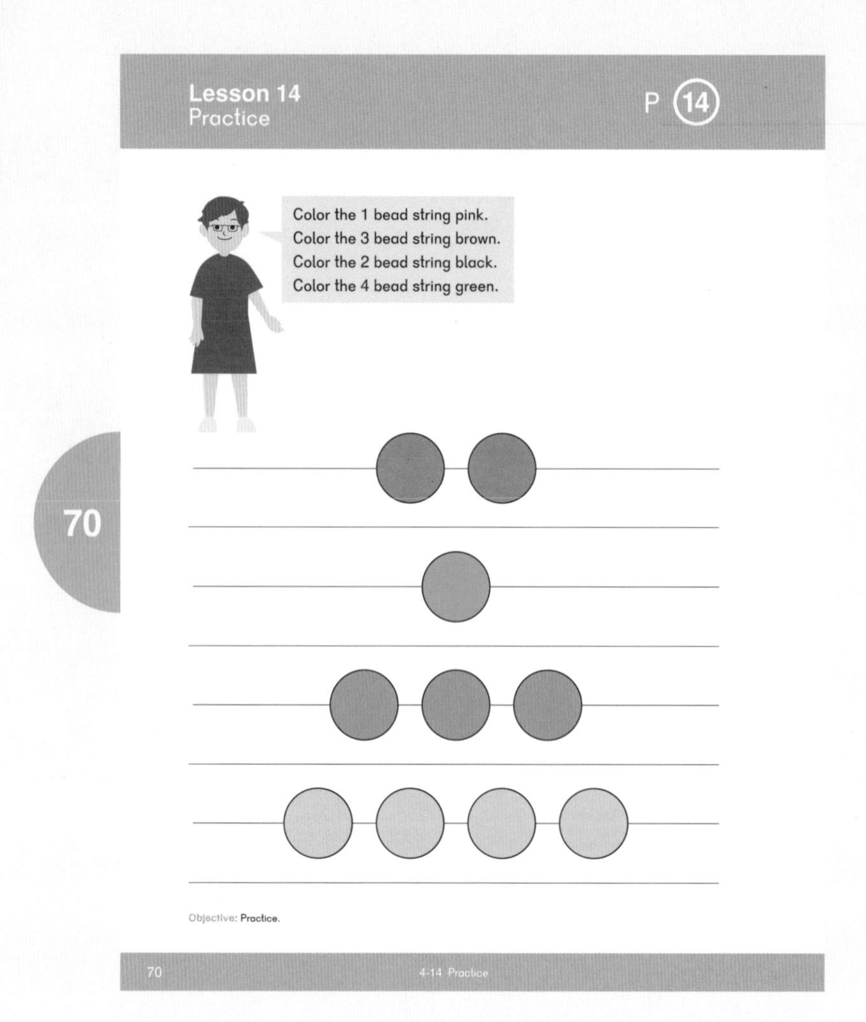

For the **Practice,** read the directions and speech bubbles on each page and have students complete the task.

Whole Group Play

Simon Says: Regular **Simon Says** rules apply. Lead students in one, two, three, four, or five movements, including showing the correct number of fingers.

Small Group Center Play

Blocks: With students working in pairs, have them take turns where one builds a tower of three, four, or five blocks and the other student tells how many blocks are used. Students then count together to verify.

Dress-up: Include clothes that babies and adults would wear, as well as brown, pink, and purple items of clothing.

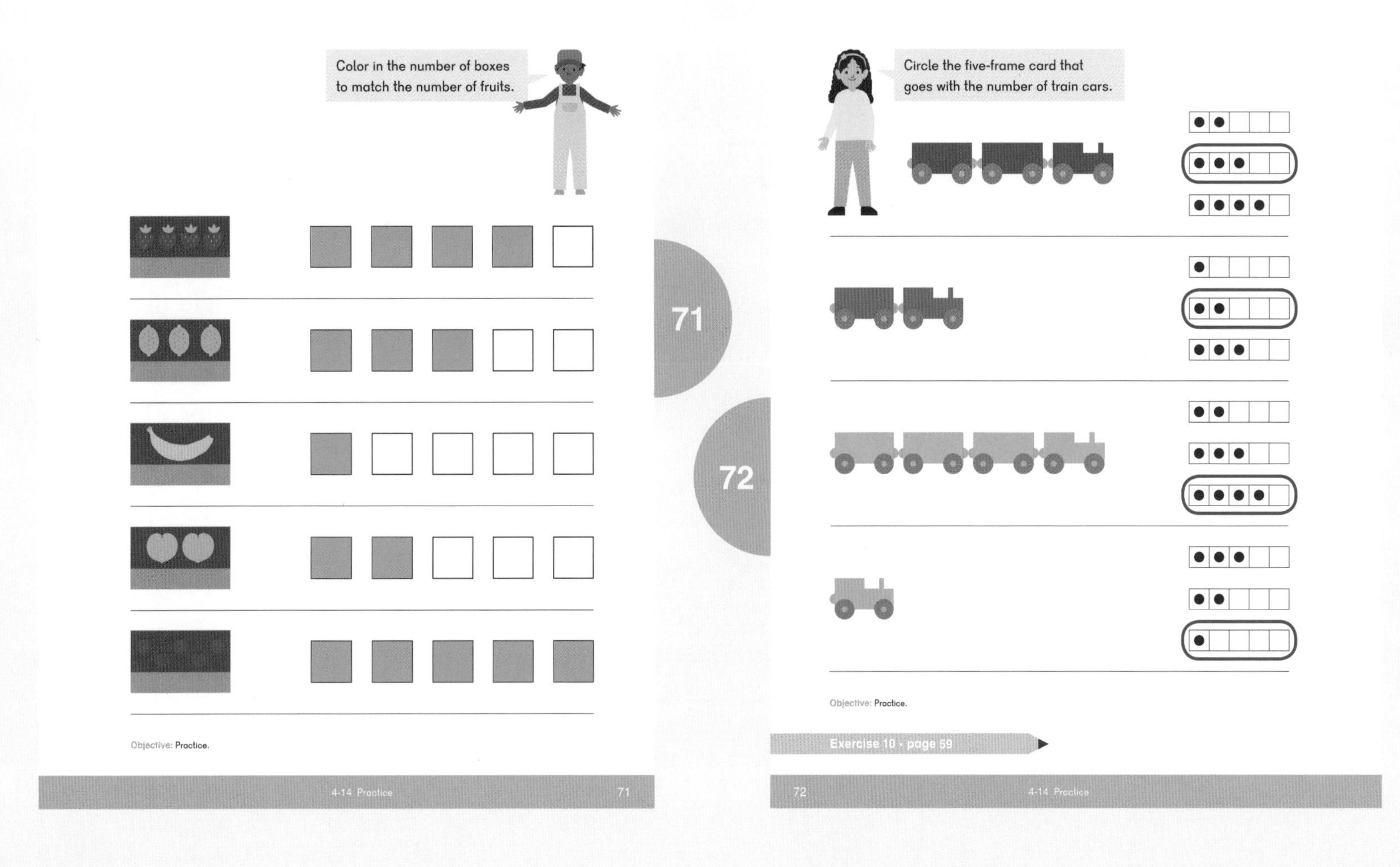

Kitchen: Include one of one item, two of another, up to five.

Art: Include five colors of paint.

Sort and Count: Provide a container holding several different types of small counters, several of each type. Have students sort the counters by type, then show a specified number of each type of counter on a work mat.

Exercise 10 • page 59

Notes

Exercise 1 • pages 39 – 40

Exercise 2 • pages 41 – 42

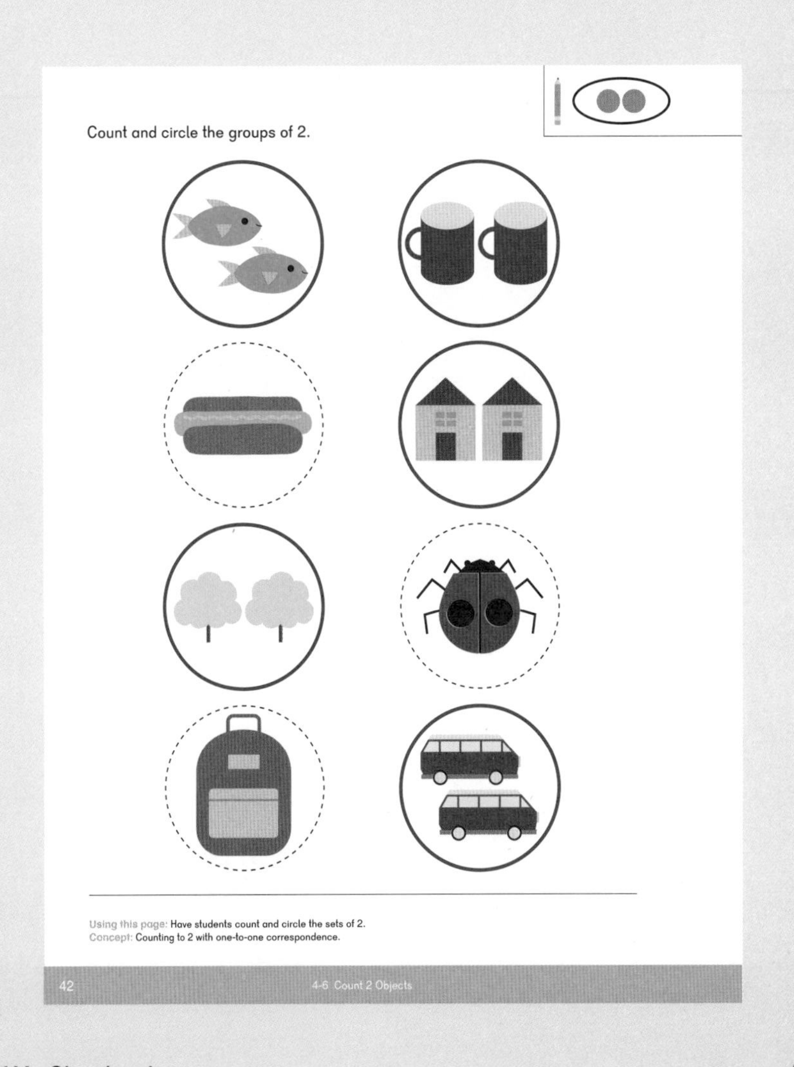

Exercise 3 • pages 43 – 44

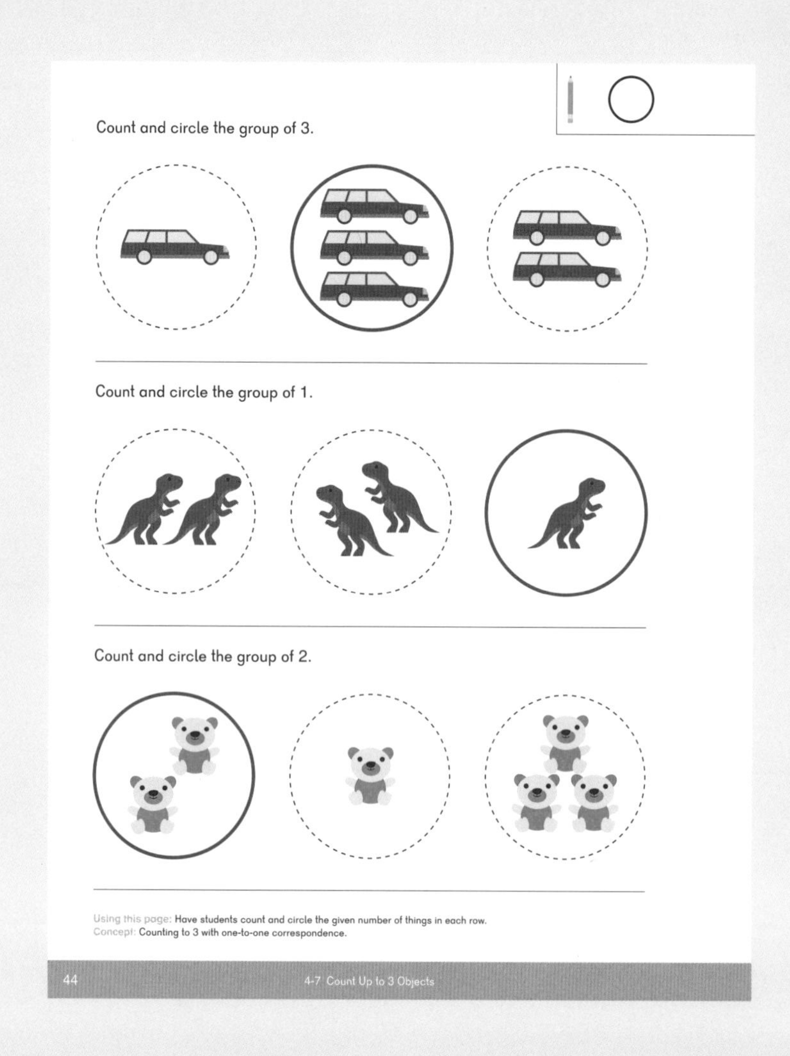

Exercise 4 • pages 45 – 48

Exercise 4

Count and circle the groups of 4.

Using this page: Have students count and circle the 3 sets of 4.
Concept: Counting to 4 with one-to-one correspondence.

4-8 Count Up to 4 Objects 45

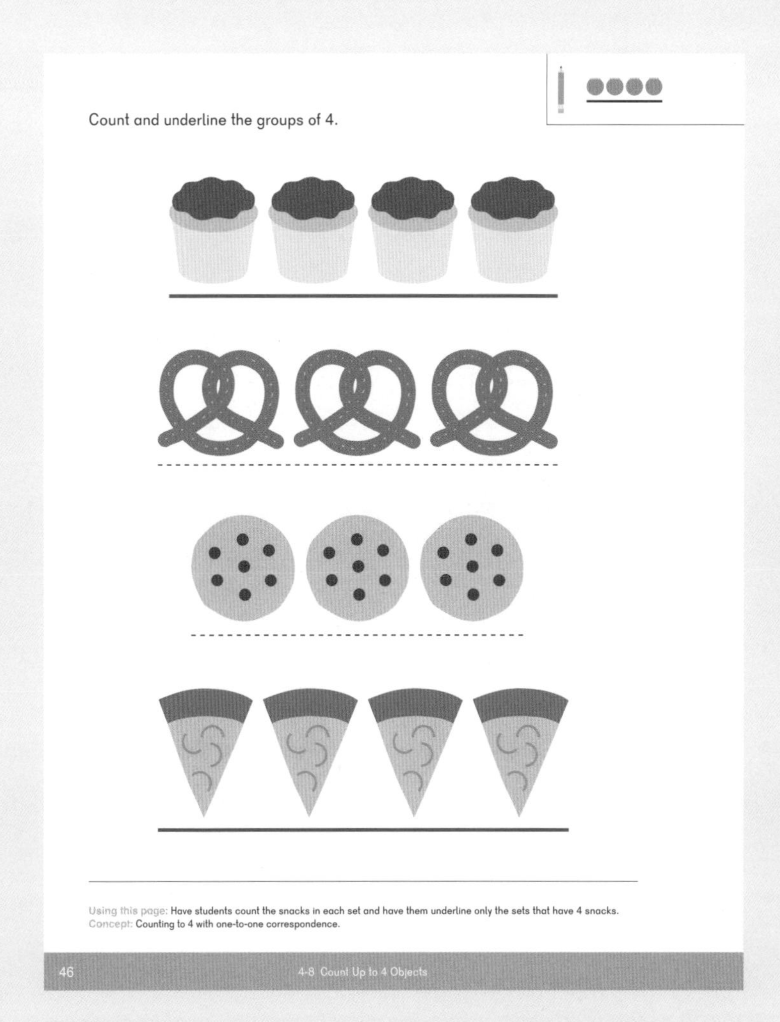
Count and underline the groups of 4.

Using this page: Have students count the snacks in each set and have them underline only the sets that have 4 snacks.
Concept: Counting to 4 with one-to-one correspondence.

46 4-8 Count Up to 4 Objects

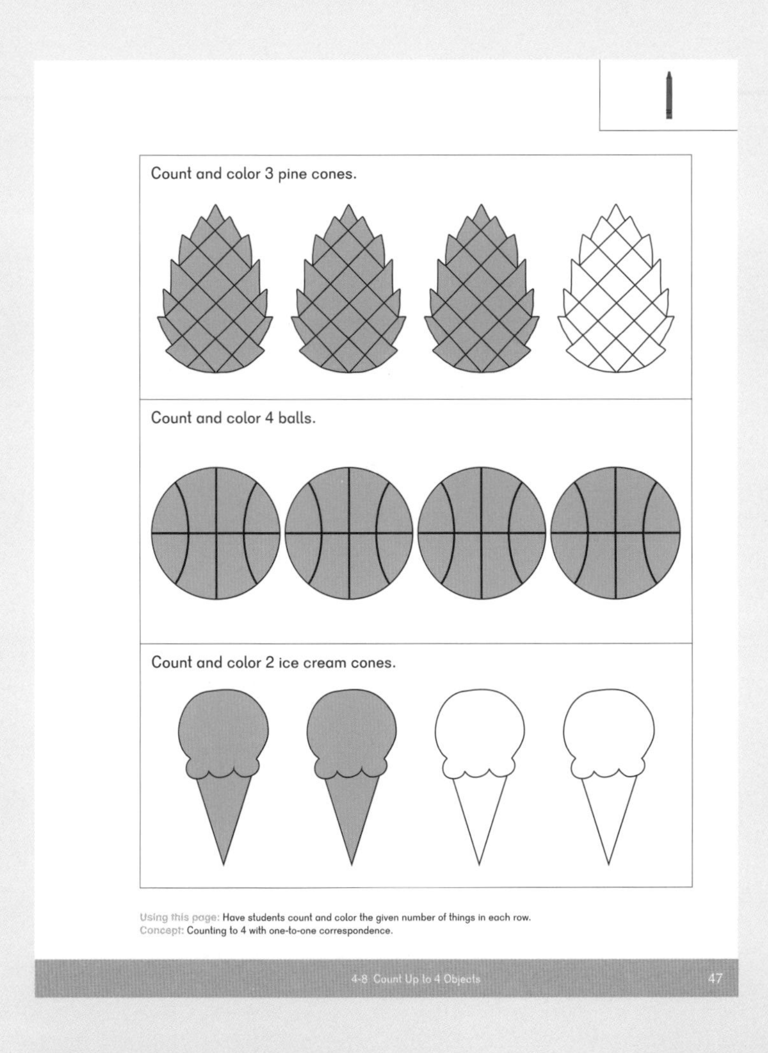
Count and color 3 pine cones.

Count and color 4 balls.

Count and color 2 ice cream cones.

Using this page: Have students count and color the given number of things in each row.
Concept: Counting to 4 with one-to-one correspondence.

4-8 Count Up to 4 Objects 47

Spot the differences and circle.

Using this page: Have students look at the 2 pictures carefully to spot the 4 differences and circle them in the bottom picture.
Concept: Counting to 4 with one-to-one correspondence.

48 4-8 Count Up to 4 Objects

Exercise 5 • pages 49 – 50

Exercise 5

Count and circle the groups of 5.

Using this page: Have students count and circle the 3 sets of 5.
Concept: Counting to 5 with one-to-one correspondence.

4-9 Count Up to 5 Objects 49

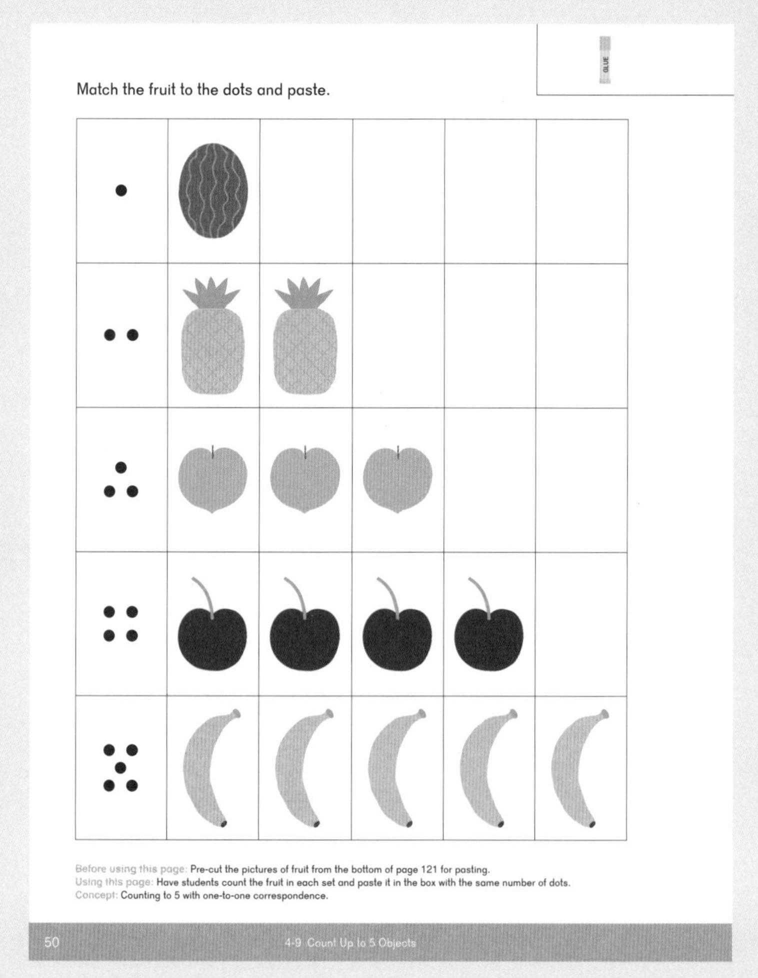

Match the fruit to the dots and paste.

Before using this page: Pre-cut the pictures of fruit from the bottom of page 121 for pasting.
Using this page: Have students count the fruit in each set and paste it in the box with the same number of dots.
Concept: Counting to 5 with one-to-one correspondence.

50 4-9 Count Up to 5 Objects

Exercise 6 • pages 51 – 52

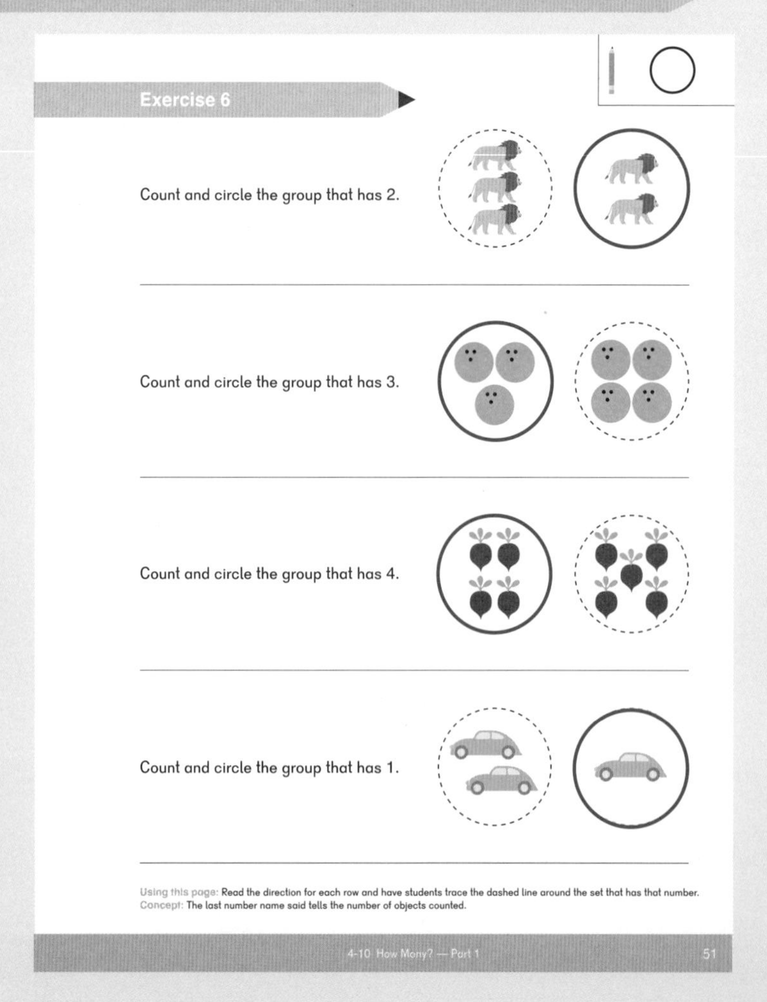

Exercise 6

Count and circle the group that has 2.

Count and circle the group that has 3.

Count and circle the group that has 4.

Count and circle the group that has 1.

Using this page: Read the direction for each row and have students trace the dashed line around the set that has that number.
Concept: The last number name said tells the number of objects counted.

4-10 How Many? — Part 1 51

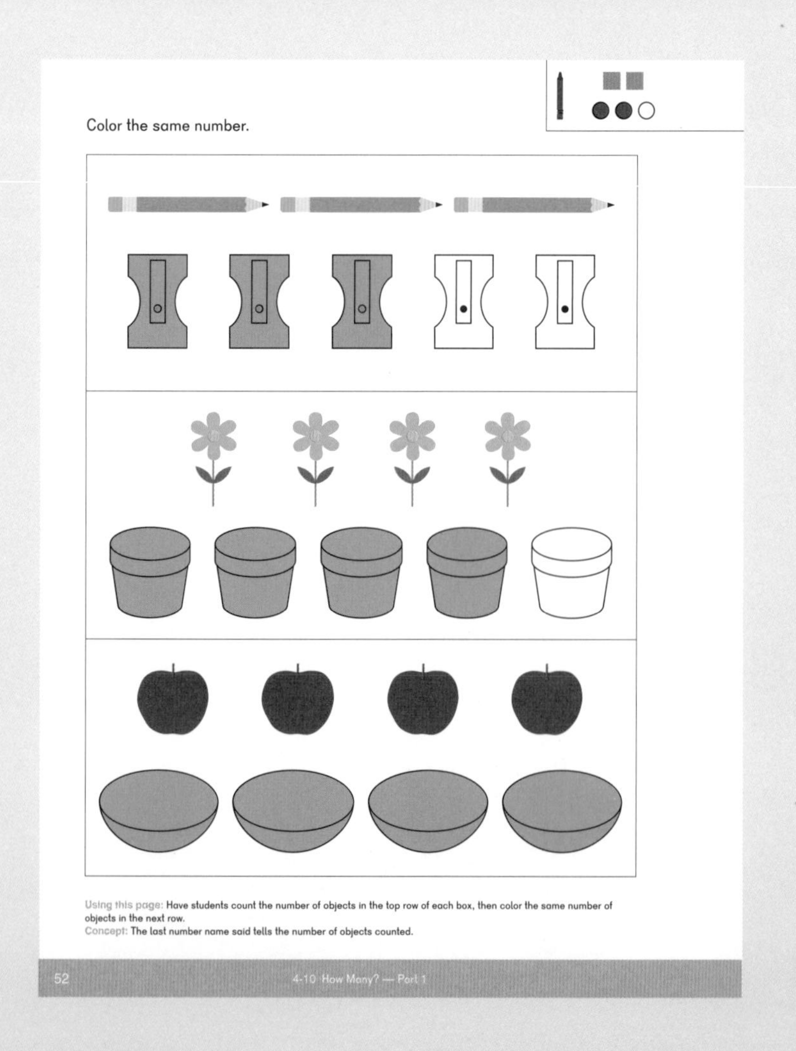

Color the same number.

Using this page: Have students count the number of objects in the top row of each box, then color the same number of objects in the next row.
Concept: The last number name said tells the number of objects counted.

52 4-10 How Many? — Part 1

Exercise 7 • pages 53 – 54

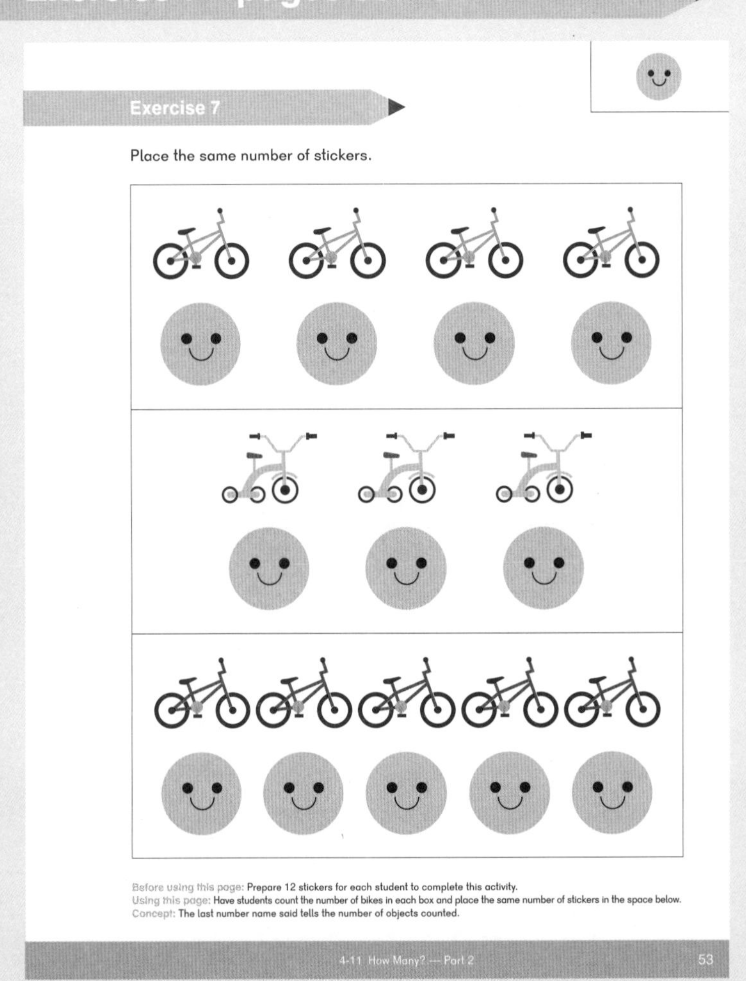

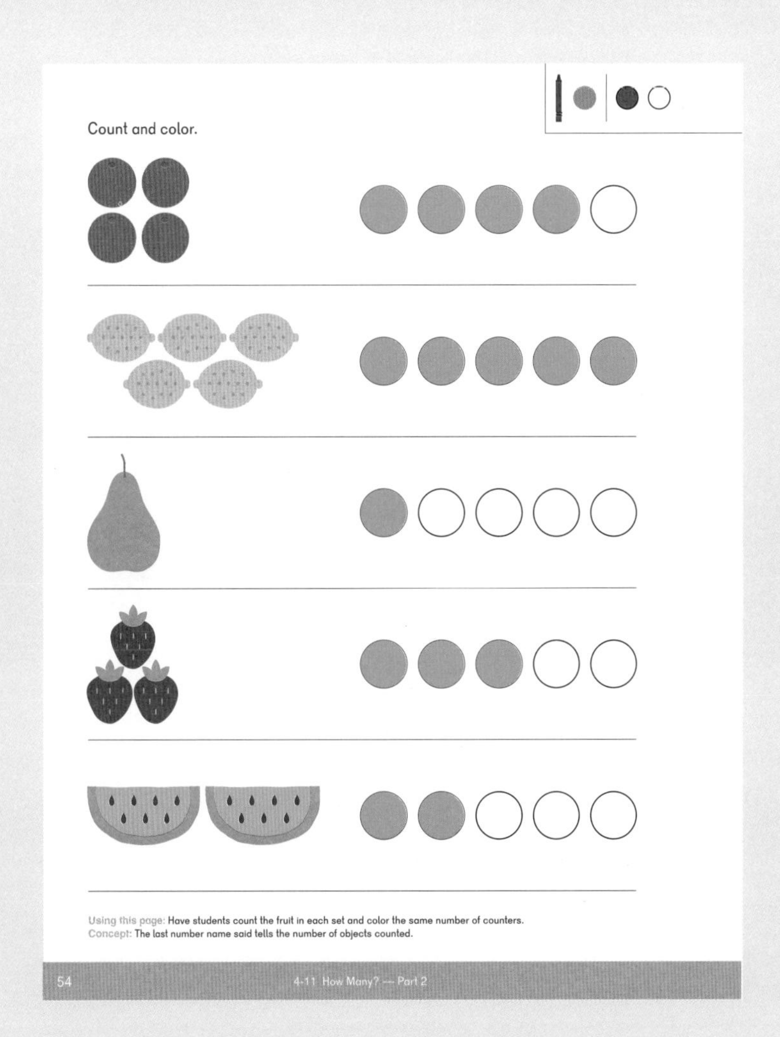

Exercise 8 • pages 55 – 56

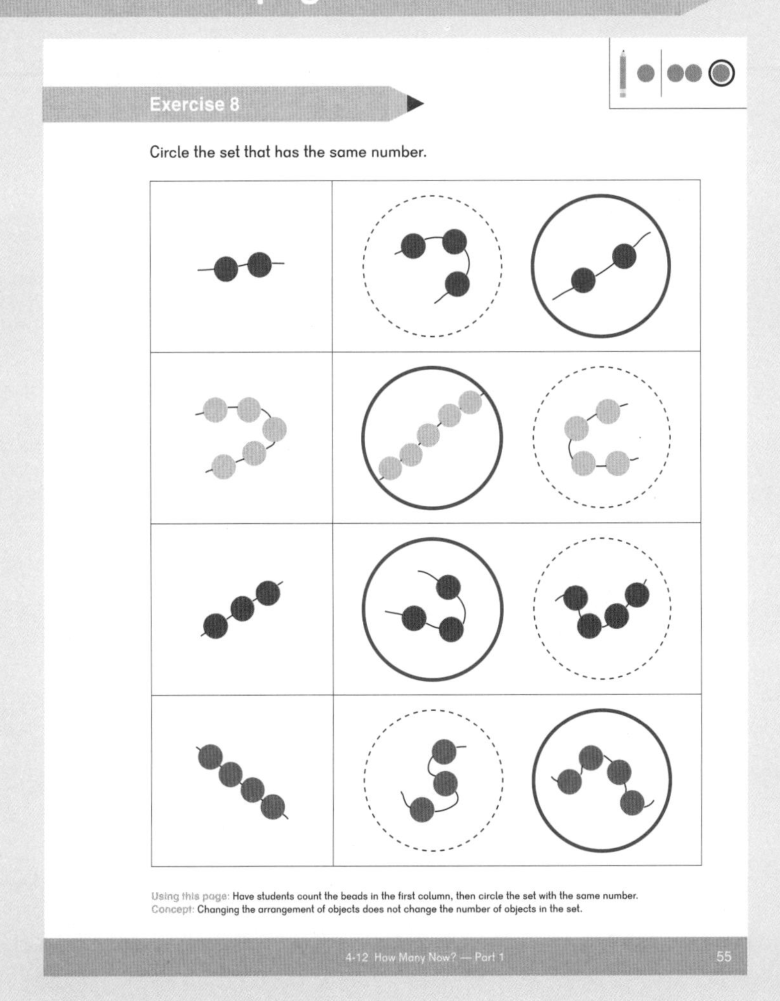

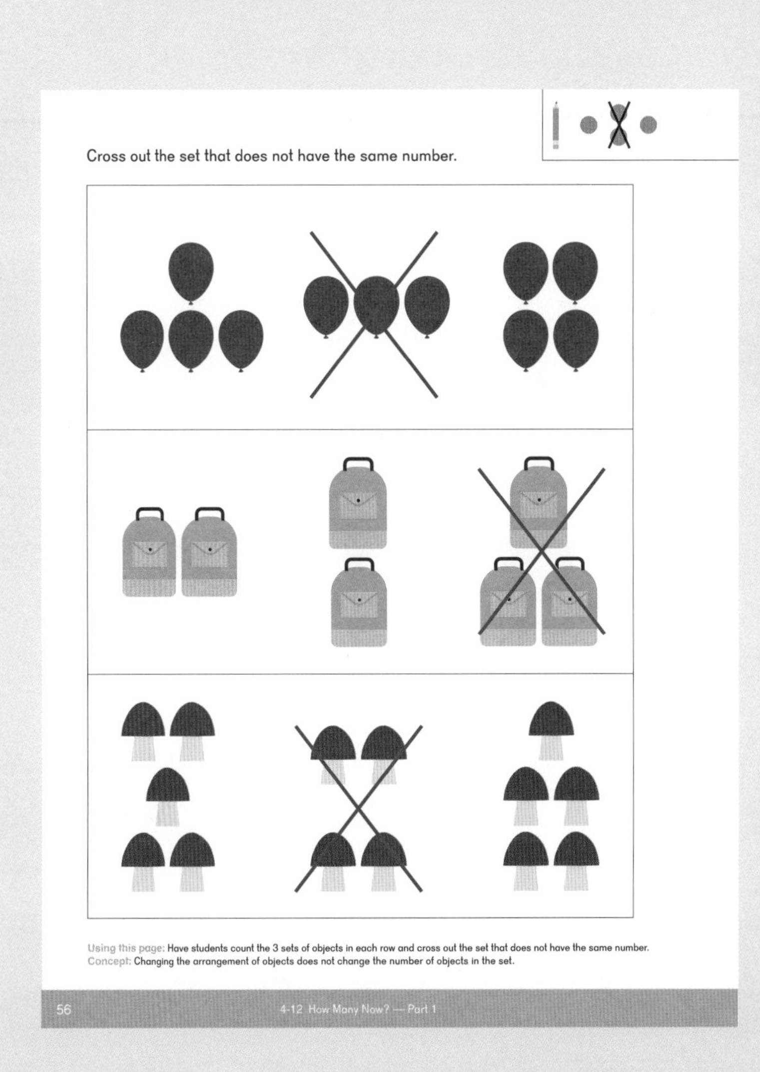

Exercise 9 • pages 57 – 58

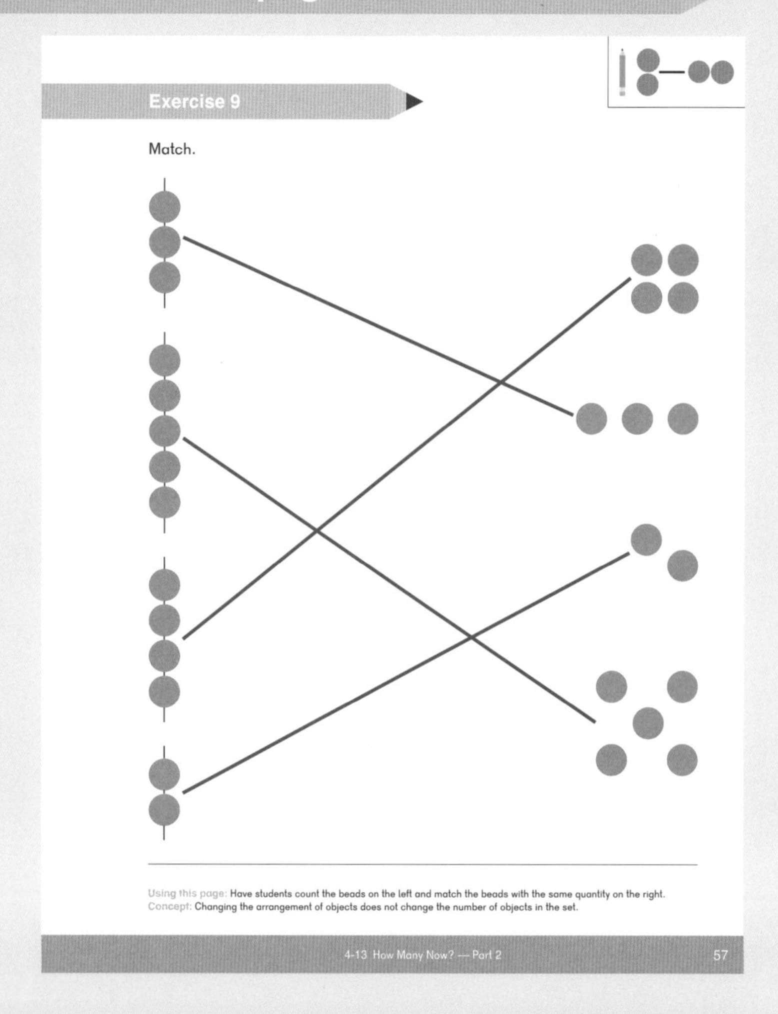

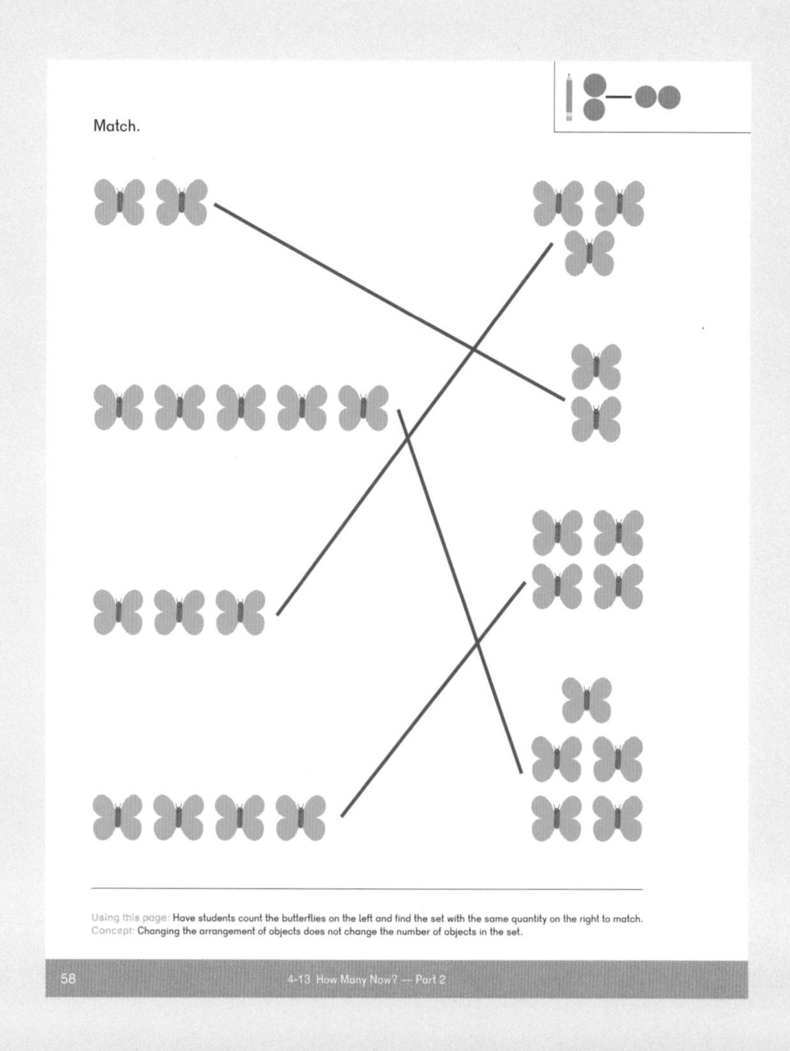

Exercise 10 • pages 59 – 60

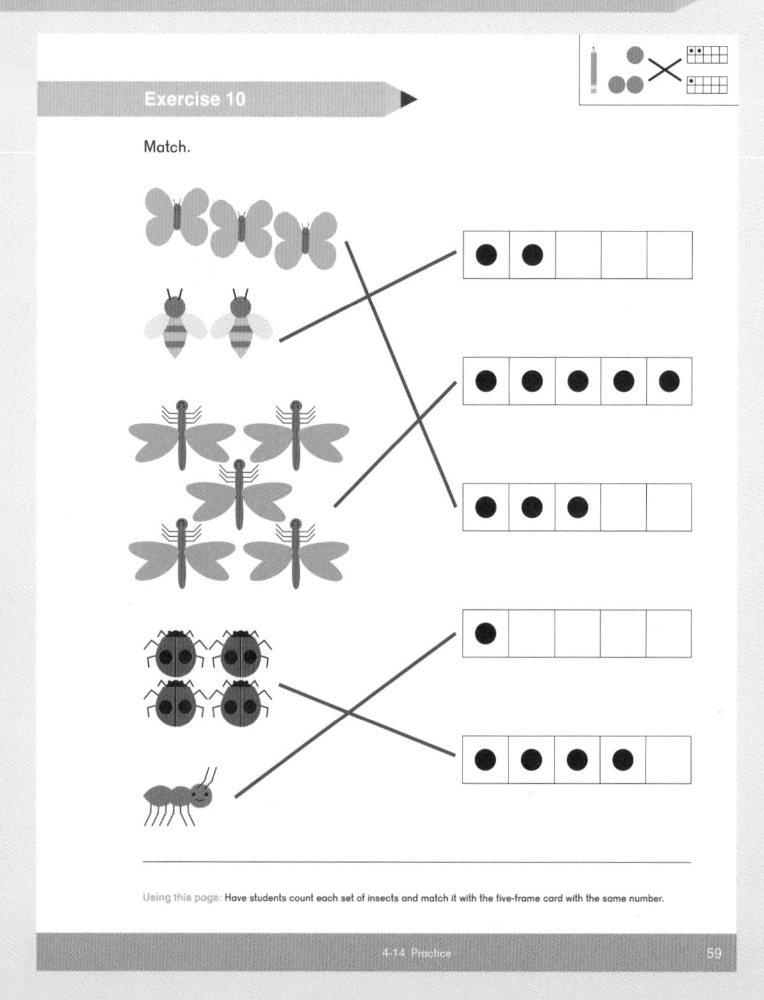

Suggested number of class periods: 8 – 9

	Lesson	Page	Resources	Objectives
	Chapter Opener	p. 111	TB: p. 73	
1	1, 2, 3	p. 112	TB: p. 74 WB: p. 61	Recognize numerals 1, 2, and 3.
2	1, 2, 3, 4, 5 — Part 1	p. 114	TB: p. 76 WB: p. 65	Recognize numerals 1 to 5.
3	1, 2, 3, 4, 5 — Part 2	p. 116	TB: p. 77 WB: p. 69	Recognize numerals 1 to 5 on a number path.
4	How Many? — Part 1	p. 118	TB: p. 78 WB: p. 71	Match a given numeral, 1 to 5, to a set containing that number of objects.
5	How Many? — Part 2	p. 120	TB: p. 80 WB: p. 73	Match a given numeral, 1 to 5, with a picture containing that number of objects.
6	How Many Do You See?	p. 122	TB: p. 82 WB: p. 75	Recognize a number of objects, up to 5, without counting.
7	How Many Do You See Now?	p. 124	TB: p. 83 WB: p. 77	Recognize that rearranging a number of objects does not change the number of objects in the set.
8	Practice	p. 126	TB: p. 86 WB: p. 79	Practice concepts introduced in this chapter.
	Workbook Solutions	p. 128		

Chapter Vocabulary

- Number path
- Number card
- Number cube
- Arrangement

In **Chapter 4: Numbers to 5 — Part 1**, students learned to count by rote, count with one-to-one correspondence, and count with cardinality. In this chapter, students will learn to recognize the numerals from 1 to 5, then to match the appropriate numeral to a set containing that number of objects.

Writing numerals will not be taught in the **Dimensions Math®** series until Kindergarten. Instead, basic and advanced counting competencies are stressed. If students in your class already know how to write the numerals, do not discourage them from doing so, but there is no need for formal instruction.

A "numeral" refers to the written symbol for a quantity, and "number" refers to the quantity. Students will call both numbers and numerals "numbers," which is acceptable at this age. Number cards showing 1 to 5 will be used extensively in this chapter.

The purpose of this chapter is for students to recognize those numerals. In the beginning, because students have learned to count objects to five and are learning to subitize, use number cards that show not only the numeral, but also a five-frame representation of the number. This type of card can be found in the **Blackline Masters** and is named **Number Cards — FFR.** Beginning with **Lesson 3: 1, 2, 3, 4, 5,** try using number cards without the five-frame representations for those students who are ready.

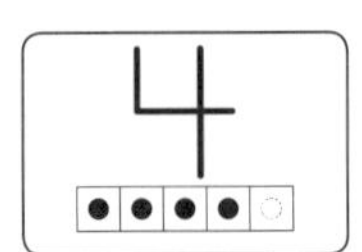

Students will be using numeral cards and numeral cubes, but will refer to them as number cards and number cubes. For that reason, they are listed as such in the materials lists for the lessons.

The **Dimensions Math®** series will use number paths, not number lines. These paths will resemble board games, hopscotch, calendars, and other familiar real world sequences of numbers. Number lines will be introduced in Grade 2.

Subitizing, the ability to instantly recognize a quantity without counting, is emphasized in this chapter. Much research has been done in recent years regarding the importance of subitizing. The conclusion is that there is a clear relationship between subitizing ability, numeracy, and general mathematics skills. Recognizing "how many" without counting will help students count on from a recognized patterned set and combine numbers from sets.[1]

Any arrangement of three on a five-frame represents three, and therefore is correct, but at the beginning, encourage the students to see that lining them up from left to right, with no gaps, is easiest for recognizing the number. Later in the chapter, students will vary the arrangements. This mathematical variability will help build subitizing skills.

Depending on your students' prior knowledge, you may choose to spend more than one day on some of the lessons in this chapter. If so, repeat the **Play** (and **Extend** if appropriate) portions of the lessons each day.

Key Points

Students arrive in Pre-Kindergarten at varying levels of experience and development. If students are struggling with lessons in this chapter, include a review center where those students can continue to experience numeral recognition as you continue to move through the curriculum.

---

[1] To learn more about subitizing, read Clements, D. H. (1999). Subitizing: What is it? Why teach it? *Teaching Children Mathematics.*

Materials

- Bags
- Baskets
- Linking cubes
- Keyboards
- Large number path (or create one using painter's tape)
- Chalk
- Bean bags
- Small counters
- Dot stickers
- Dice, small and large
- Teddy bear counters
- Number cubes
- Index cards
- Stickers
- Large-grid graph paper
- Square tiles of any rigid material
- Straws
- Medicine droppers
- Sponges
- Play food
- Play kitchen items
- Bases (t-ball bases or paper plates, taped down)
- Balls (to play t-ball)

Note: Materials for Activities will be listed in detail in each lesson.

Blackline Masters

- Number Cards
- Number Cards — Large
- Number Cards FFR — Small
- Number Cards FFR — Large
- Individual Number Path
- Five-frame Cards 1 to 5
- Blank Five-frame
- Picture Cards
- Dot Cards

Storybooks

- *Goodnight Gorilla* by Peggy Rathmann

Optional Snacks

- Pretzel sticks
- Apple slices
- Cream cheese
- Cheese sticks
- Gelatin cut into number shapes 1 through 5
- Graham crackers
- Animal crackers
- Celery sticks
- Raisins
- Crackers

Letters Home

- Chapter 5 Letter

Notes

Lesson Materials

- *Goodnight Gorilla* by Peggy Rathmann
- Number Cards FFR — Small (BLM), with each number printed on a different color of paper

Explore

Have students discuss any experiences they have had at a zoo and describe the animals they have seen in zoos.

Read *Goodnight Gorilla* aloud, focusing on any animals of which there are only one, two, or three. When you find such animals, have students count them with you as you put your finger on each one.

Learn

Have students look at page 73 and identify the animals on the page. Have them count each animal by putting a finger on each, then tell you, for example, "There is one camel." Ask them why the five-frames shown are where they are on the page.

Give each student a number card showing a five-frame. There should be two 1s, two 2s, etc. The students' task will be to find other students with the matching number. Then have the pair of students pick an animal from the story and act it out. Have other students guess which animal the acting student chose.

Objective

- Recognize numerals 1, 2, and 3.

Lesson Materials

- Number Cards (BLM), 1 set per student
- Number Cards FFR — Small (BLM) 1, 2, and 3, 1 set per student
- Number Cards — Large (BLM) 1, 2, and 3
- Bags containing 3 linking cubes, 1 bag per student
- Optional snack: Pretzel sticks or apple slices with numbers 1, 2, or 3 written on them with cream cheese

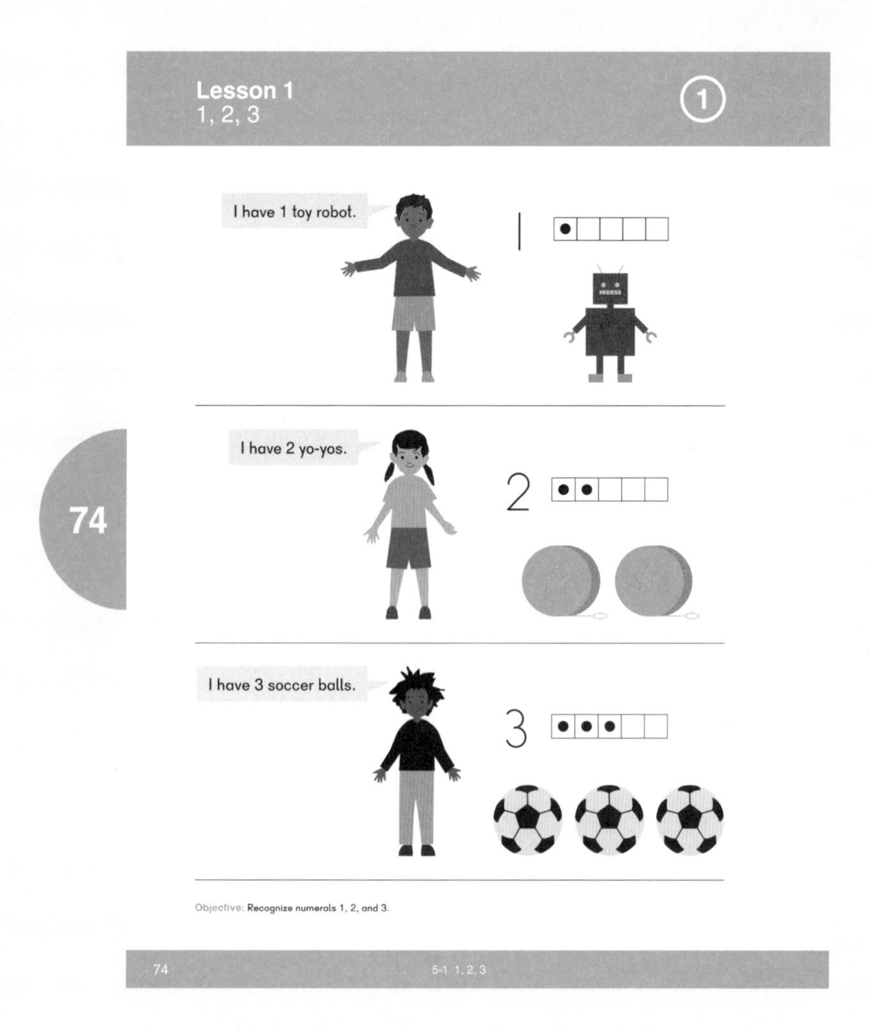

Explore

Give each student a set of Number Cards FFR — Small (BLM). Hold up a Number Card FFR — Large (BLM) and have students call out the number. This is a quick, informal assessment of students' background knowledge.

Give each student a bag of linking cubes. Hold up a Number Card (BLM) 1 and one object. Have students show the matching card. Then have them show that many linking cubes. Repeat for 2 and 3.

Learn

Have students look at page 74 and identify the type and number of toys each friend has.

Show students a number card 1 while holding up one object and saying, "One." Repeat for 2 and 3. Next, hold up a number card, 1, 2, or 3, and the correct number of objects, and have students call out the correct number.

Hold up a number card 1 and have students describe the number. Repeat for 2 and 3. Students may notice that a 1 has a straight line, a 2 is curved at the top and has a straight line across the bottom, and a 3 has no straight lines.

Hold up the three number cards at the same time and have students say what is alike and what is different about the three numerals.

Have students look at page 75. Read Mei's directions aloud and have them complete the task.

Whole Group Play

Body Part Numbers: Have students stand side-by-side facing the front of the classroom. Call out a body part (head, shoulders, knees, toes, eyes, ears, mouth, or nose) and have students put their fingers on the body part called. Show a Number Card — Large (BLM) 1, 2, or 3, and have students touch the body part called that number of times.

Materials: Number Cards — Large (BLM) 1, 2, and 3

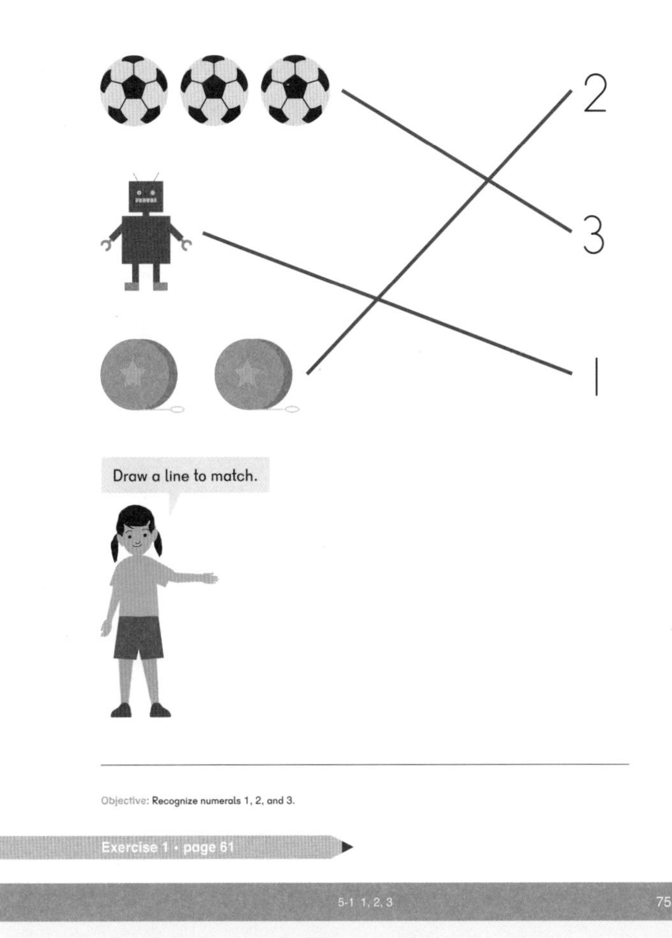

Head, Shoulders, Knees, and Toes (VR): Sing the song and have students touch the correct body part as they sing.

Head, shoulders, knees and toes
Knees and toes
Head, shoulders, knees and toes
Knees and toes
Eyes and ears and mouth and nose
Head, shoulders, knees and toes
Knees and toes

Small Group Center Play

Sort: Include one of some objects, two of others, three of others, and number cards 1, 2, 3. Have students sort the objects in different ways, pairing a number card with each group. If the technology is available, take pictures of some of the interesting sorts and put them up on the Math Wall.

Kitchen: Provide kitchen items and/or play food representing each quantity up to the numbers 1, 2, and 3. For example, you might provide one saucepan, two plates, and three play apples.

Art: Include three colors of paint, including brown, and art paper on which you have written large numerals 1, 2, and 3. Have students trace the numerals with paint.

75

Exercise 1 • page 61

Extend Play

What Sounds Different?: Have students play with a keyboard, noticing the difference in sound between an F and an F#. Play the first two measures of "Heads, Shoulders, Knees, and Toes" for them and see if they can find the rest of the notes to the song.

Head, shoulders, knees and toes
G A G F# G E G G G

Materials: Keyboards

Sing It Backwards: Sing the song from the lesson backwards:

Toes, knees, shoulders head, shoulders head
(Repeat)
Nose and mouth and ears and eyes,
toes knees shoulders head, shoulders head

Objective

- Recognize numerals 1 to 5.

Lesson Materials

- Small object sets or pictures of sets containing 4 of one type and 5 of a different type
- Number Cards – Large (BLM) 1 to 5
- Number Cards FFR – Large (BLM)
- Number Cards (BLM), 1 set per student
- Linking cubes – 5 cubes per set, 1 set per student
- Optional snack: Cheese sticks in the shape of 4s

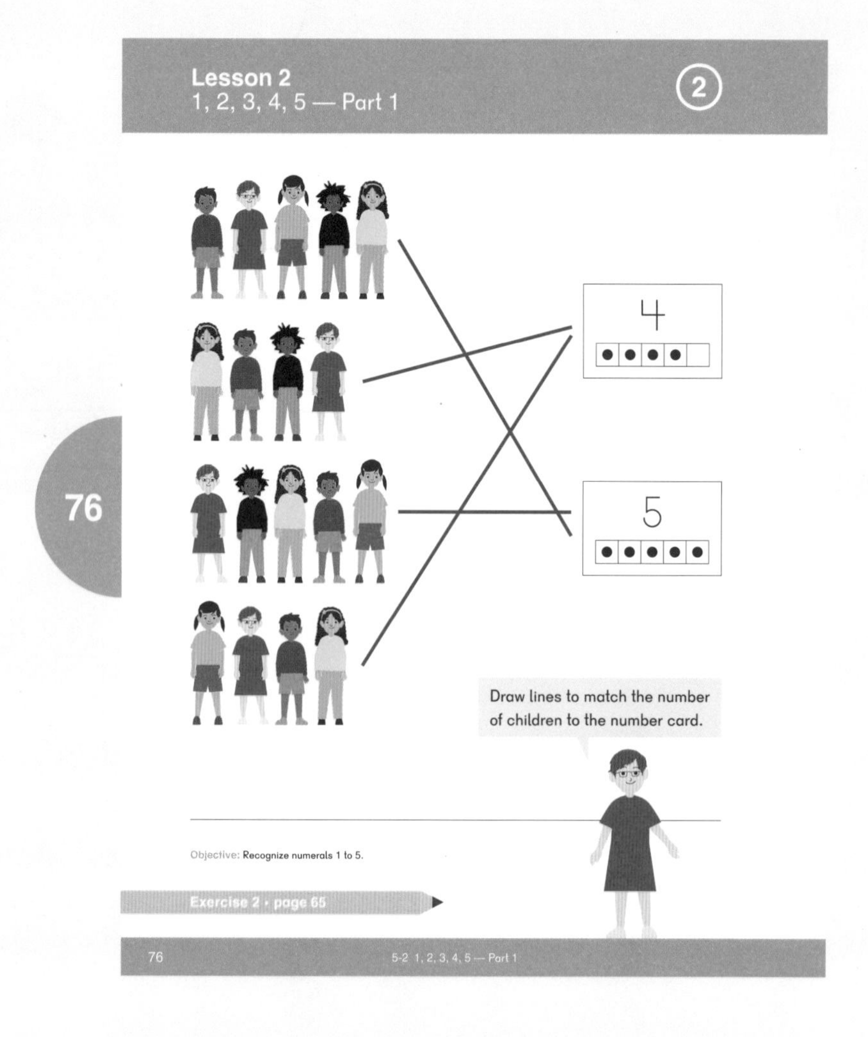

Prior to the beginning of this lesson, “hide” sets of four and five objects, or pictures of sets of four and five objects, around the room.

Explore

Send four or five students at a time to find sets of four objects or pictures of sets of four objects and bring them back to the group. After all students have gathered their finds, discuss as a group. Repeat for five.

Note: Post the numerals 1 to 5 and refer to them often from now on.

Bring student focus back to the whole group. Show them a Number Card FFR — Large (BLM) with a 4 while holding up four objects and saying, “Four.” Repeat for five. Hold up a 4 or 5 Number Card — Large (BLM) and the correct number of objects and have students tell you the correct number.

Hold up a Number Card (BLM) 4 and have students describe the numeral. Repeat for 5. Students may notice that a 4 has three straight lines and that a 5 has two straight lines and a curvy part. Then show all five numerals at once and have students talk about how 4 and 5 are alike and different from 1, 2, and 3. You may choose to say, “My friend keeps mixing up the number 3 and the number 5. How could I help him remember which is which?”

Give each student a set of Number Cards FFR — Small (BLM) 1 to 5 and five linking cubes. Call out a number, 1 to 5, and have students show that many linking cubes and the correct number card.

Hold up a number card 1 to 5 and call out a movement (clap, touch toes, pat head, etc.). Students’ task will be to do that movement the number of times shown on the card you hold up. Repeat with other cards and movements.

Learn

Have students look at page 76. Ask them to identify the missing friends in the second row (Mei) and the fourth row (Dion). Read Emma's direction and have students complete the task.

Whole Group Play

Run to Your Number!: Set up bases outside with a Number Card (BLM) 1 to 5 on each. There will be three bases for each of the numbers. Have students run to a base with the correct number card on it as you call numbers 1 to 5.

Materials: 3 sets of Number Cards — Large (BLM) 1 to 5, 15 bases

Small Group Center Play

Sort: Include one of some objects, two of others, etc., to five, and number cards 1 to 5. Have students sort the objects in different ways, pairing a number card with each group. If the technology is available, take pictures of some of the interesting sorts and put them up on the Math Wall.

Kitchen: Include objects in quantities of one, two, three, four, and five. Give students number cards three, four, and five and have them match the objects to the number cards.

Sponge Art: Include five colors of paint and sponges of different sizes. Have them dip a sponge in one color and transfer the paint to paper. Have them dip a different sponge into a different color two times, then another sponge three times, transferring paint to paper each time. Repeat for four and five.

Exercise 2 • page 65

Objective

- Recognize numerals 1 to 5 on a number path.

Lesson Materials

- Large number path for classroom floor similar to the one shown on textbook page 77
- Individual Number Path (BLM) 1 to 5, 1 per student
- Number Cards FFR — Small (BLM) 1 to 5, 1 set per student
- Optional snack: Gelatin cut into numbers 1 to 5

Explore

Give each student an Individual Number Path (BLM). Set up a large number path on the floor. Ask students what they notice about the path. How is it different from number cards?

Some students may say that they can see all of the numbers at once. Some may say that the numbers are connected to each other. Some may even notice something about how the numbers increase from left to right.

Give each student a set of Number Cards FFR — Small (BLM) 1 to 5.

Have a student stand facing the number path. Think of a number and describe the number with a clue to have the student hop on that number. For example, to get the student to hop on 1, you could say, "Hop to the number of noses I have." Each student takes two turns. Meanwhile, the other students use their fingers to "hop" on their individual number paths.

Learn

Have students look at page 77 and say what Dion is doing. Read Dion's comments. Have students say why Dion is where he is in each case.

Whole Group Play

Straight Line Hopscotch: Draw several number paths with chalk outside, or using painter's tape inside. Divide students into groups and give each group a beanbag. First student tosses a bag onto a numeral and hops to it, counting each hop on the way. For example, if the beanbag lands on 4, the student hops (or jumps) and everybody counts, "1, 2, 3, 4." Hopping student then picks up the beanbag and passes it to the next student.

Materials: Painter's tape, chalk, beanbags

Small Group Center Play

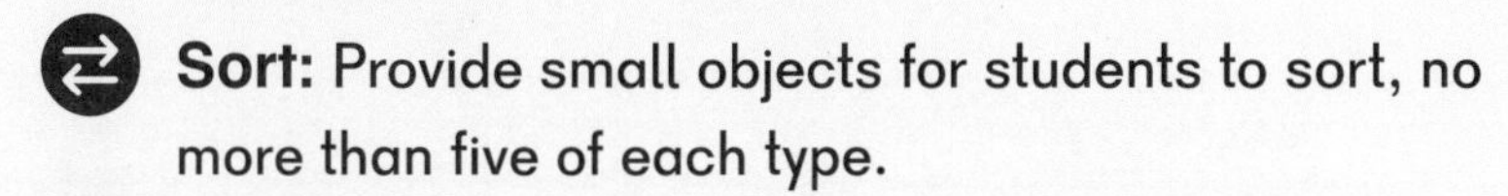

Sort: Provide small objects for students to sort, no more than five of each type.

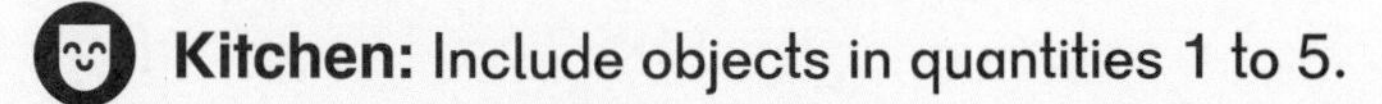

Kitchen: Include objects in quantities 1 to 5.

Paint the Number: Include two colors of paint and paper with large numerals 4 and 5 on them. Students' task will be to paint the 4 one color and paint the 5 a different color by putting four drops of paint on the 4 and five drops of paint on the 5.

Exercise 3 • page 69

Extend Play

Describe That Number: Give each student a set of Number Cards (BLM) 1 to 5 and have them verbally describe each numeral in as much detail as possible. They may either describe the numerals to you or into a recorder.

After you have had time to listen to and comment on students' descriptive language, have students work in pairs. One student will verbally describe a numeral and the other student will name the numeral based on the description.

Materials: Number Cards (BLM) 1 to 5 for each student

Objective

- Match a given numeral, 1 to 5, to a set containing that number of objects.

Lesson Materials

- Five-frame Cards (BLM) 1 to 5
- Number Cards (BLM) 1 to 5, for each student
- 5 wooden blocks or other similar objects
- Optional snack: Graham crackers with numbers 1 to 5 written on them with cream cheese

Explore

Hold Number Cards (BLM) 1 to 5, mixed-up, facedown. Have three or four students at a time pick a card then find a quantity of objects in the classroom that matches the number. As some students are hunting, lead the other students in patterned clapping while saying, “They're going on a treasure hunt.” As students find their sets of objects, have them return to the group and share their “finds.”

Give each student a set of number cards. Hold up a wooden block and have students hold up the number card (1) that matches. Repeat with two, three, four, and five blocks. Next, repeat the activity using Five-frame Cards (BLM) 1 to 5 instead of blocks. Mix up the numbers. Point out to students that the five-frame cards always show dots colored starting on the left side. Help them remember which is the left side by remembering which pinkie they start with when counting their fingers.

Learn

Have students look at page 78 and discuss Emma's comment. Ask students which objects they recently counted in the classroom.

Have students look at page 79 and identify the object. Remind students that Alex likes to play t-ball. Read Alex's directions and have them complete the task.

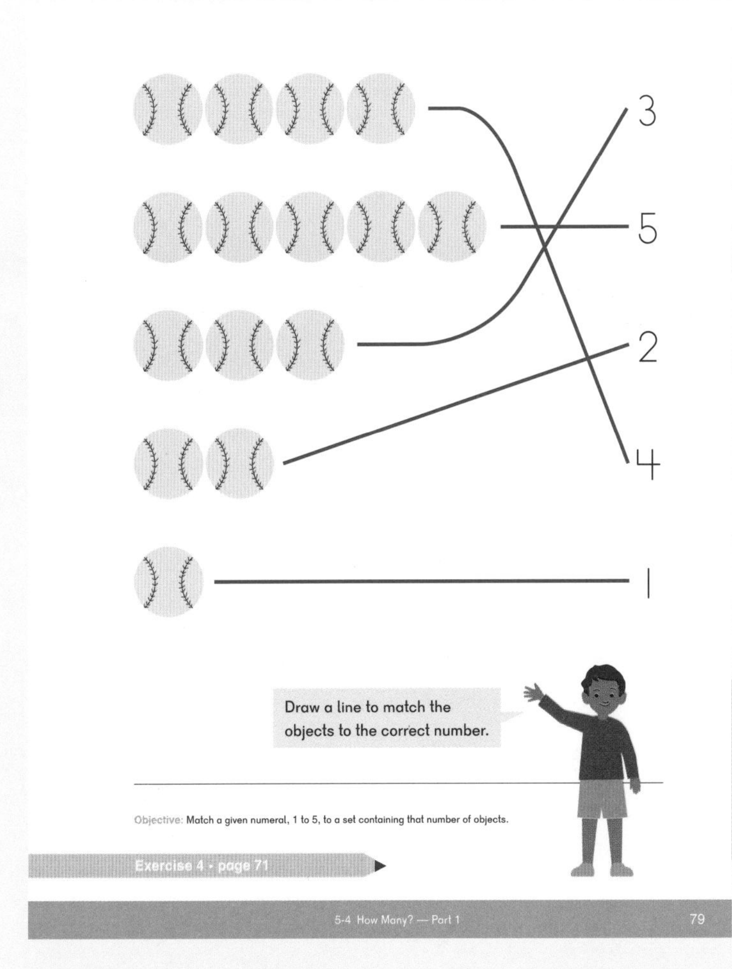

79

Small Group Center Play

Sort: Set out building blocks of different sizes, colors, etc., and Number Cards (BLM) 1 to 5. Have students build towers using similar blocks for different numbers 1 to 5 and place the card with the tower.

Kitchen: Include objects in quantities of 1 to 5.

Sticker Cards: Have students make dot cards showing 1 to 5. Then have them post their cards on the math wall or chart paper showing the correct numeral.

Match: Provide Number Cards (BLM) 1 to 5 and linking cube trains 1 to 5, and have students match those representing the same number. For example, the student would match a train of five linking cubes with the number card showing 5.

Exercise 4 • page 71

Whole Group Play

Do You Remember?: Pick out three Number Cards (BLM) and show them to the students, one at a time. As you show each card, place it facedown in front of you. When all three cards are facedown, point to one of the cards and have students predict which one it is. Turn the card up. Repeat with the other two cards. Repeat with other number cards. This game will help students develop visual memory skills.

Materials: Number Cards (BLM) 1 to 5

T-ball Toss: Have students play catch using t-balls.

Materials: T-balls or other balls

Extend Learn

Do You Remember? (with a twist): After the three Number Cards (BLM) are placed facedown, move their positions while telling students to watch you move them, before having students predict each card. You can also play with more than three cards.

Materials: Number Cards (BLM) 1 to 5

Objective

- Match a given numeral, 1 to 5, with a picture containing that number of objects.

Lesson Materials

- Picture Cards (BLM)
- Small counters, 5 per student
- Blank Five-frame (BLM), 1 per pair of students
- Number Cards (BLM) 1 to 5, 1 set per student
- Optional snack: Animal crackers

80

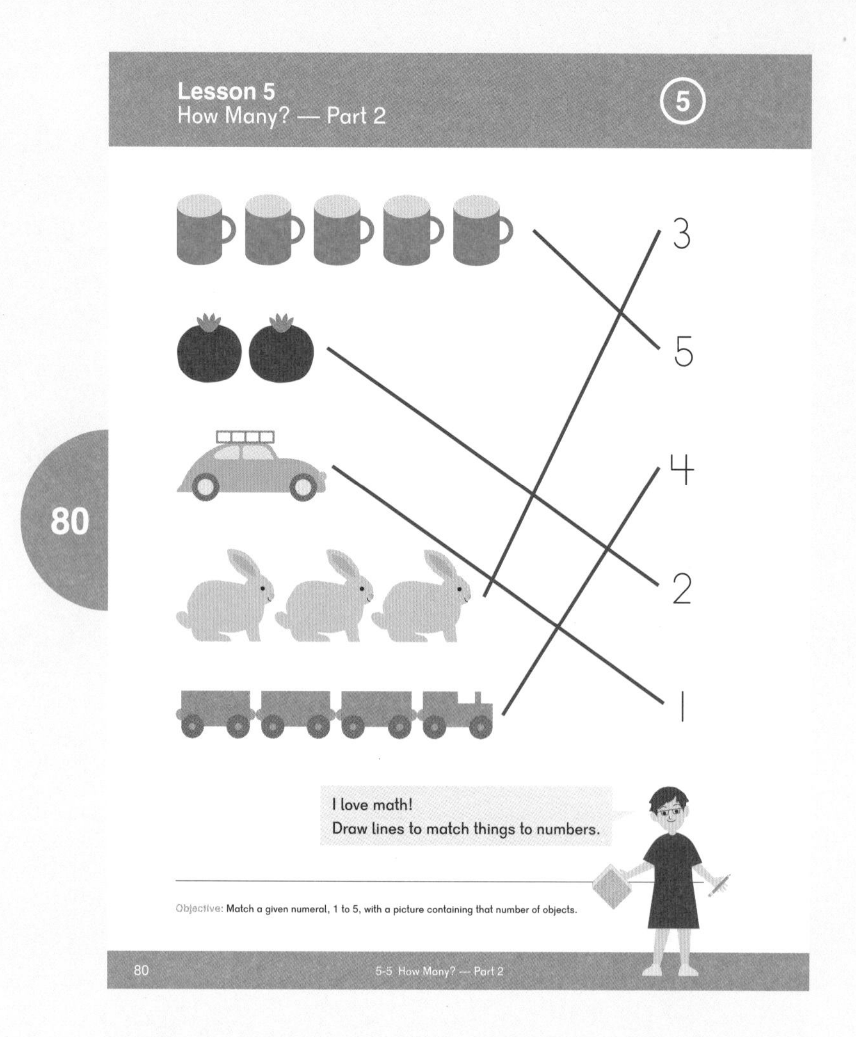

Explore

Give each pair of students Picture Cards (BLM), a Blank Five-frame (BLM), counters, and Number Cards (BLM) 1 to 5. One student chooses a picture card and uses the five-frame cards and counters to show the number of objects on the card. The other student finds the matching number card. Have students change roles several times.

Learn

Have students look at page 80 and identify the objects. Tell them that Emma loves math. Ask them to say what they do in math that is fun. Read Emma's direction and have them complete the task.

Have students look at page 81. Read Sofia's comment and direction. Ask them if they like to draw. Have them use a finger, then a pencil, to trace the square drawn on the page, then complete the task.

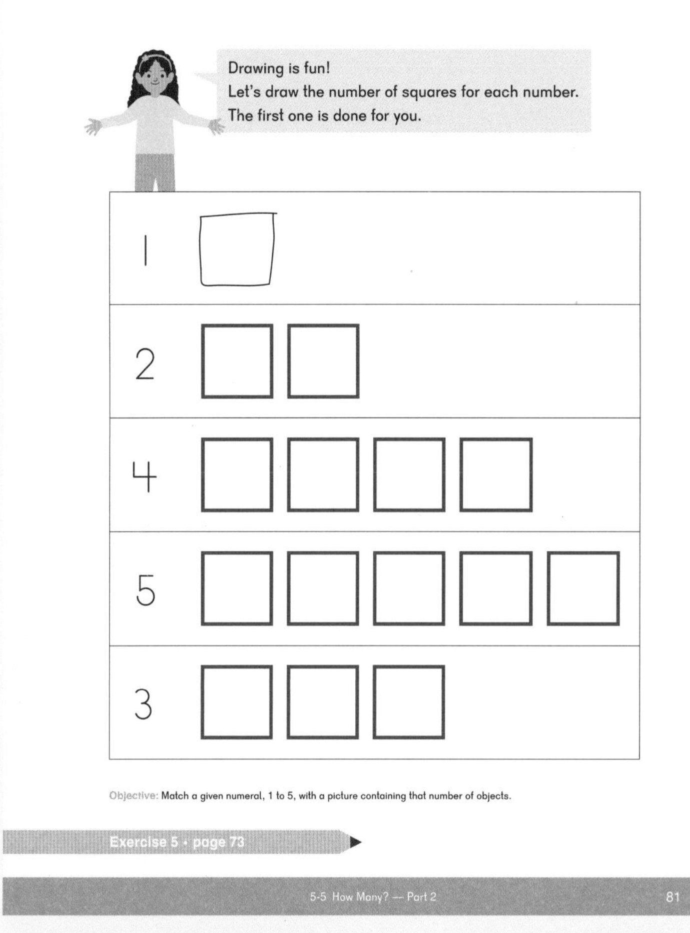

81

Small Group Center Play

Blocks: Set out building blocks and have students build structures of their choice using similar blocks for different numbers 1 to 5.

Going on a Picnic: Provide baskets or bags and play food, up to 5 of any one type of food. Have students pick two Number Cards (BLM) 1 to 5 from a pile of facedown cards and pack a picnic using those numbers of different types of food.

Sticker Cards: Give students index cards and dot stickers, and have them make dot cards using stickers showing 1 to 5. Then have them post their cards on the math wall or chart paper showing the correct numeral.

Match: Provide Number Cards (BLM) 1 to 5 and Dot Cards (BLM) 1 to 5, and have students match those representing the same number.

Exercise 5 • page 73

Whole Group Play

Straight Line Hopscotch: Repeat from **Lesson 3: 1, 2, 3, 4, 5 — Part 2**. If necessary, use painter's tape on the classroom floor and have students whisper count the numbers in their groups.

Materials: Painter's tape, chalk, beanbags

Extend Play

Dot Card Flash: Flash dot cards created by students and have them identify the numbers shown.

Materials: Dot cards created by students in centers

Objective

- Recognize a number of objects, up to 5, without counting.

Lesson Materials

- Blank Five-frame (BLM), 1 per student
- Five-frame Cards (BLM) 1 to 5
- Linking cubes or teddy bear counters of different colors
- Number cubes with 1 to 5 showing 1 to 5 only (cover the 6 with tape)
- Number Cards (BLM) 1 to 5 for each student
- Dot Cards (BLM) 1 to 5 for each student
- Optional snack: Large crackers with die patterns for 1 to 5 on them made with cream cheese

Explore

Hold up Five-frame Cards (BLM) 1 to 5 and have students tell you how many dots they see on each card without counting.

Give each student three counters and a Blank Five-frame Card (BLM). Hold up a Picture Card (BLM) showing up to three objects in different arrangements, and have students show you how many they see by representing the number of objects on the picture card with counters on their five-frame cards. Show the card again and point to each object on the card as the students and you count them together. Repeat with several different picture cards.

After students are confident with subitizing, give each of them a set of Number Cards (BLM) 1 to 5. Rather than having them represent the number of objects with counters, have them hold up the appropriate number card when you flash picture cards and/or five-frame cards at them. For students not yet confident, continue to flash five-frame cards and Dot Cards (BLM) showing one, two, and three before moving on.

Roll a large die showing 1 to 5 only and have students say how many they see.

Give each pair of students a number cube or die showing 1 to 5 dots only. Tell them to take turns rolling the cube. Both students will call out the number of dots ("pips") they see as quickly as possible.

Learn

Have students look at page 82 and say the number of dots they see on the faces of Mei's die.

Whole Group Play

Picture Cards: Have students create picture cards using index cards and three, four, and five stickers in all different arrangements on each card. Have them flash their picture cards and have other students identify how many stickers they see.

Materials: Index cards, stickers

Small Group Center Play

Blocks: Set out building blocks and opaque bags. Students will work in pairs. One student will close his or her eyes. The other students will count a number of blocks, two to five, and put them in the bag. The first student will pour out the blocks and quickly say how many there are without counting.

Dress-Up: Students roll a die and use that many items of clothing and/or accessories.

Sticker Cards: Include dice showing 1 to 5 only, circle stickers, and index cards. Have students copy dice formations of 1 to 5 using stickers.

Match: Provide dice 1 to 5 and Number Cards (BLM) 1 to 5, and have students match those representing the same number.

Exercise 6 • page 75

Extend Learn

Roll and Hop: Have students work in pairs and give materials to each pair. Tell students that one student is to roll the number cube. Both students are to identify the number of dots showing on the number cube as quickly as possible. Then the student who didn't roll the die is to count the squares and move the counter the correct number of spaces. Players alternate roles until the counter goes off of the strip of paper.

Materials: Number cube showing 1 to 5 only, small counter, strip of large-grid graph paper showing 20 sections

Lesson 7 How Many Do You See Now?

Objective

- Recognize that rearranging a number of objects does not change the number of objects in the set.

Lesson Materials

- Dot Cards (BLM) showing 3, 4, and 5 dots in different arrangements
- Number Cards (BLM) 1 to 5, 1 set per student
- Optional snack: Celery sticks filled with cream cheese and 1 to 5 raisins

Explore

Play **Dot Card Flash** starting with Dot Cards (BLM) showing three in different arrangements. Repeat with cards showing four and five. At first, have students call out the numbers they see.

Give each student a set of Number Cards (BLM) 1 to 5. Play **Dot Card Flash** again having student hold up the correct number card rather than calling out the numbers.

Learn

Have students look at page 83. Read Sofia's comment aloud. Have them say if the arrangements of three remind them of anything. Some students might say that the top arrangement reminds them of eyes and a nose. Some may say that the second arrangement looks like the corners of a triangle.

Have students look at page 84 and identify the arrangements that show one, then two, then three, then four dots. Read Sofia's direction and have them complete the task.

Have students look at page 85 and identify the arrangements that show two, then three, then four, then five dots. Read Emma's direction and have them complete the task.

Whole Group Play

Arranging 4: Give each student four square tiles. Have them arrange the tiles in as many different ways as they can. Tell them that one side of each tile must completely line up with one side of another tile in their arrangements.

Materials: 4 square tiles per student

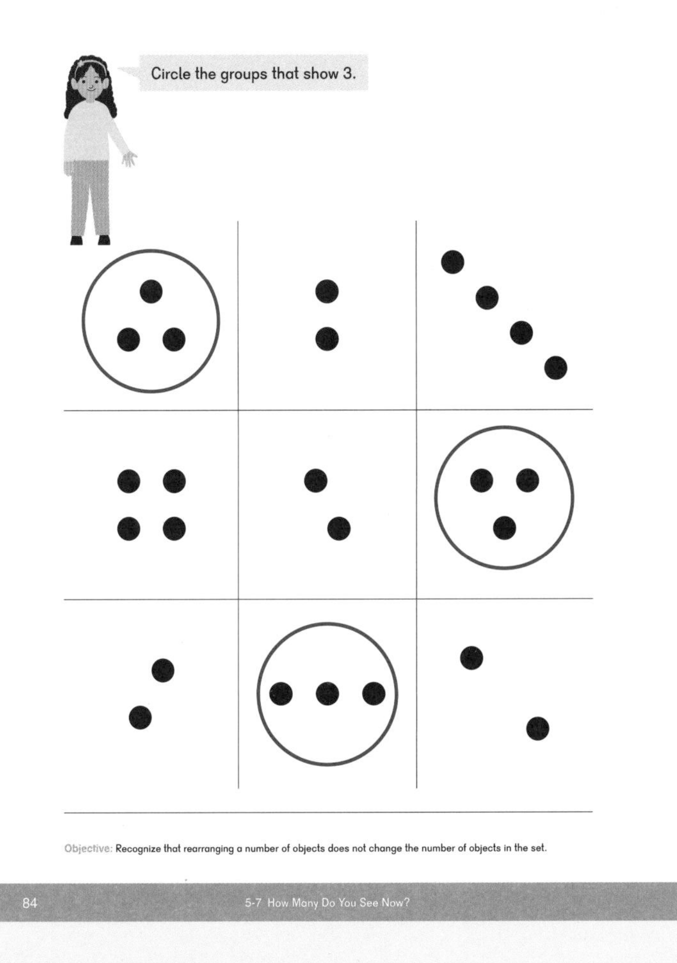

Small Group Center Play

Blocks: Have students arrange three, four, or five blocks in different ways.

Chef's Kitchen: Provide large plates and play food items. Have students arrange three, four, or five items of food on a plate in different ways.

Breath Art: Provide art paper, 5 colors of thin paint in small containers, medicine droppers, and straws for students. Have them place the directed number of paint drops on their art paper and then create art by blowing air through the straws to move the paint around the paper.

Match: Include Dot Cards (BLM), Number Cards (BLM) and Five-frame Cards (BLM) for numbers 1 to 5. A match is one of each type of card with the same number.

Exercise 7 • page 77

Extend Learn

Arranging 5: Give each student five square tiles. Have them arrange the tiles in as many different ways as they can. Tell them that each tile must touch at least one other tile in their arrangements. If the technology is available, take pictures of student work and put them up on the Math Wall.

Materials: 5 square tiles per student

Objective

- Practice concepts introduced in this chapter.

Lesson Materials

- Optional snack: Gelatin cut into numbers 1 to 5

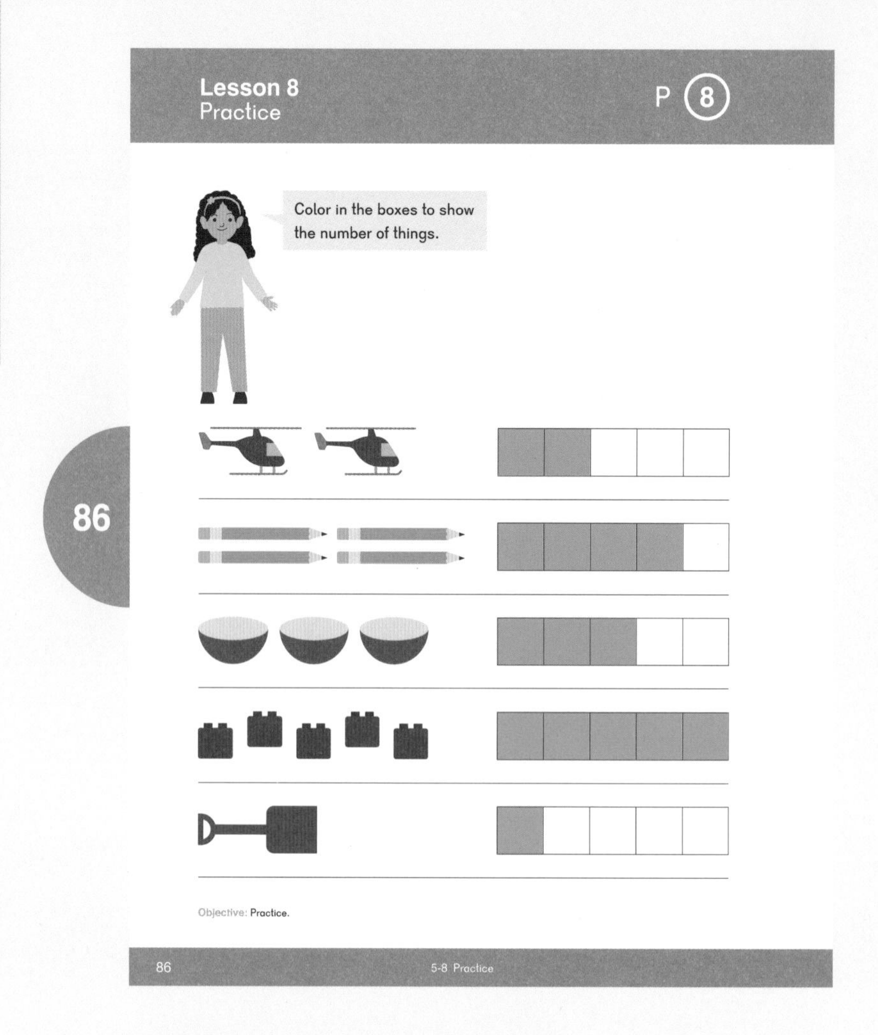

For the **Practice**, read the directions and speech bubbles on each page and have students complete the tasks.

Whole Group Play

Number Treasure Hunt: While students are completing the **Practice**, post the sticky notes around the classroom. Have four or five students go on a hunt at the same time. Give clues to the number you want them to find. Clues could be, for example, "I'm looking for the sticky note that shows the number of eyes I have," or, "I'm looking for the sticky note that shows the number of fingers I have on my left hand." While that group of students is hunting, lead the other students in a clapping pattern saying, "They're going on a treasure hunt."

Materials: Sticky notes with the numbers 1 to 5 written on them

Small Group Center Play

Blocks: Have students arrange three, four, or five blocks in different ways.

Grocery Store: Provide Number Cards (BLM) 1 to 5, bags, and items to buy. One student plays the store worker. Other students hold up a number card and call out a food item. Store worker counts out that many items and places them in a bag for each customer.

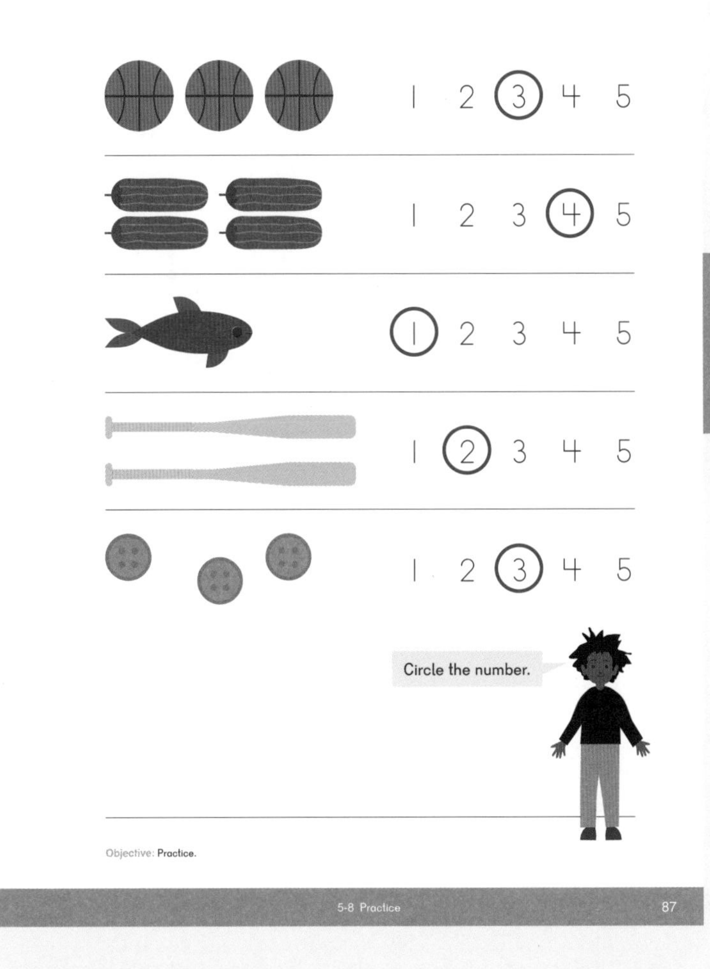

87

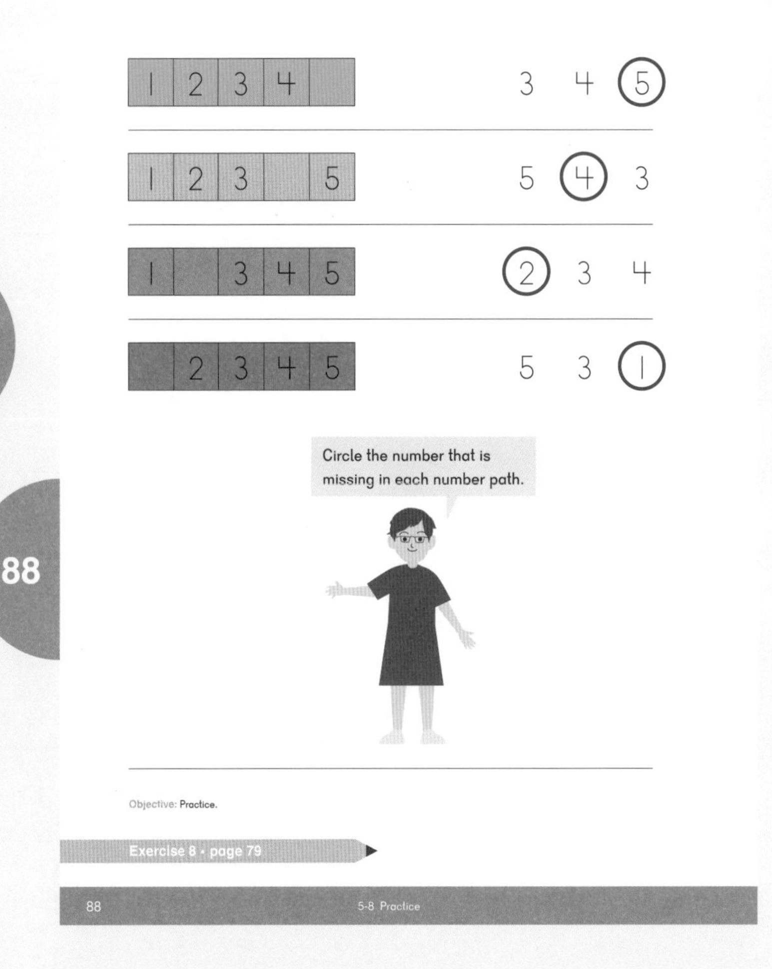

88

Breath Art: Provide art paper, 5 colors of thin paint in small containers, medicine droppers, and straws for students. Have them place the directed number of paint drops on their art paper and then create art by blowing air through the straws to move the paint around the paper.

Match: Include Dot Cards (BLM), Number Cards (BLM), and Five-frame Cards (BLM), for numbers 1 to 5. A match is one of each type of card.

Note: Students have not specifically done an activity like the one shown on page 88. They have, however, worked and played with number paths throughout this chapter. This **Practice** activity is included to see if students can extend their understanding.

Exercise 8 • page 79

Extend Play

Make More Clues: Have students create clues for the **Number Treasure Hunt.**

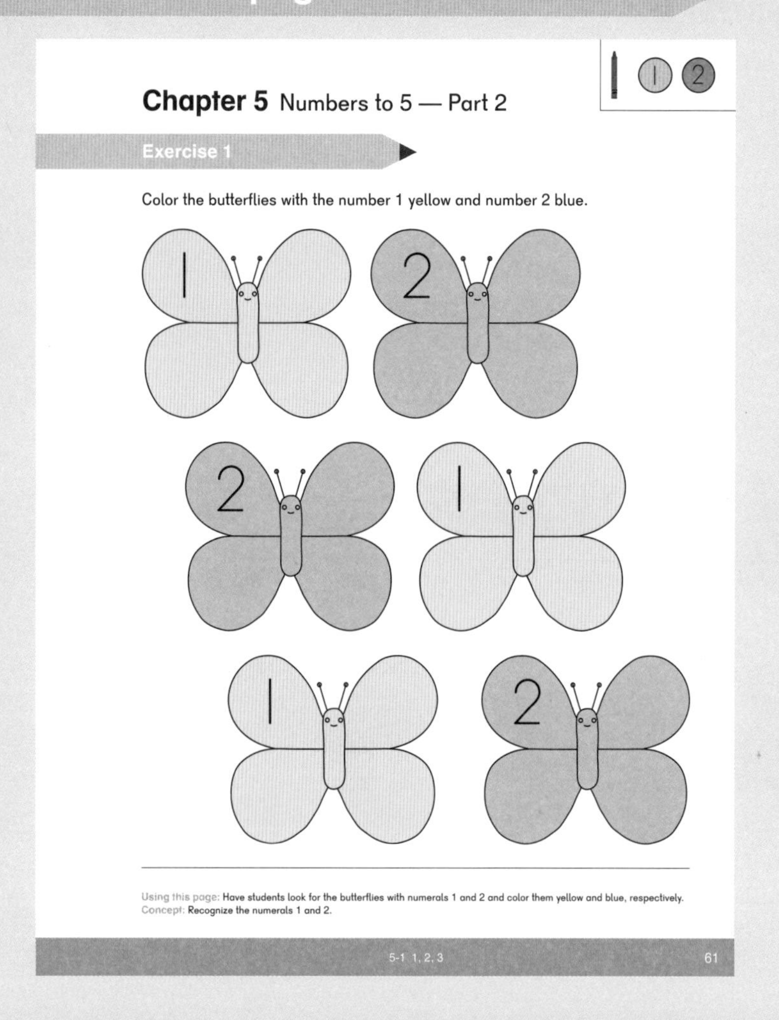
Chapter 5 Numbers to 5 — Part 2
Exercise 1
Color the butterflies with the number 1 yellow and number 2 blue.
1
2
2
1
1
2
Using this page: Have students look for the butterflies with numerals 1 and 2 and color them yellow and blue, respectively.
Concept: Recognize the numerals 1 and 2.
5-1 1, 2, 3
61

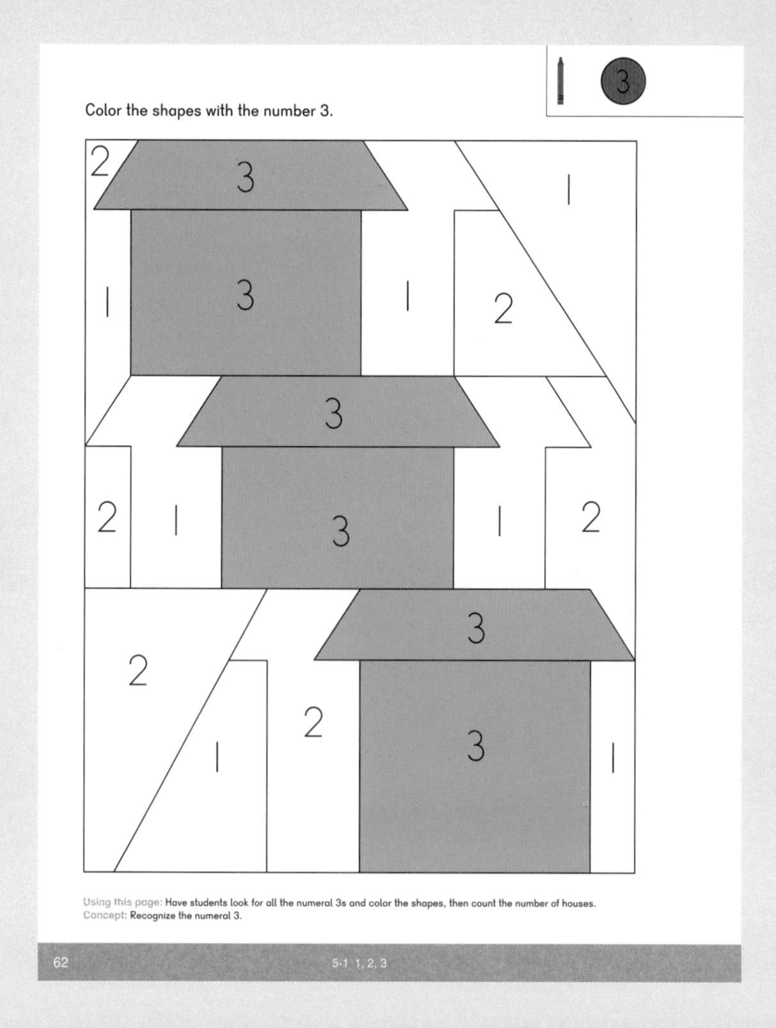
Color the shapes with the number 3.
Using this page: Have students look for all the numeral 3s and color the shapes, then count the number of houses.
Concept: Recognize the numeral 3.
62
5-1 1, 2, 3

Match.
1
3
2
Using this page: Have students match each numeral to the correct set of animals.
Concept: Recognize the numerals 1 to 3.
5-1 1, 2, 3
63

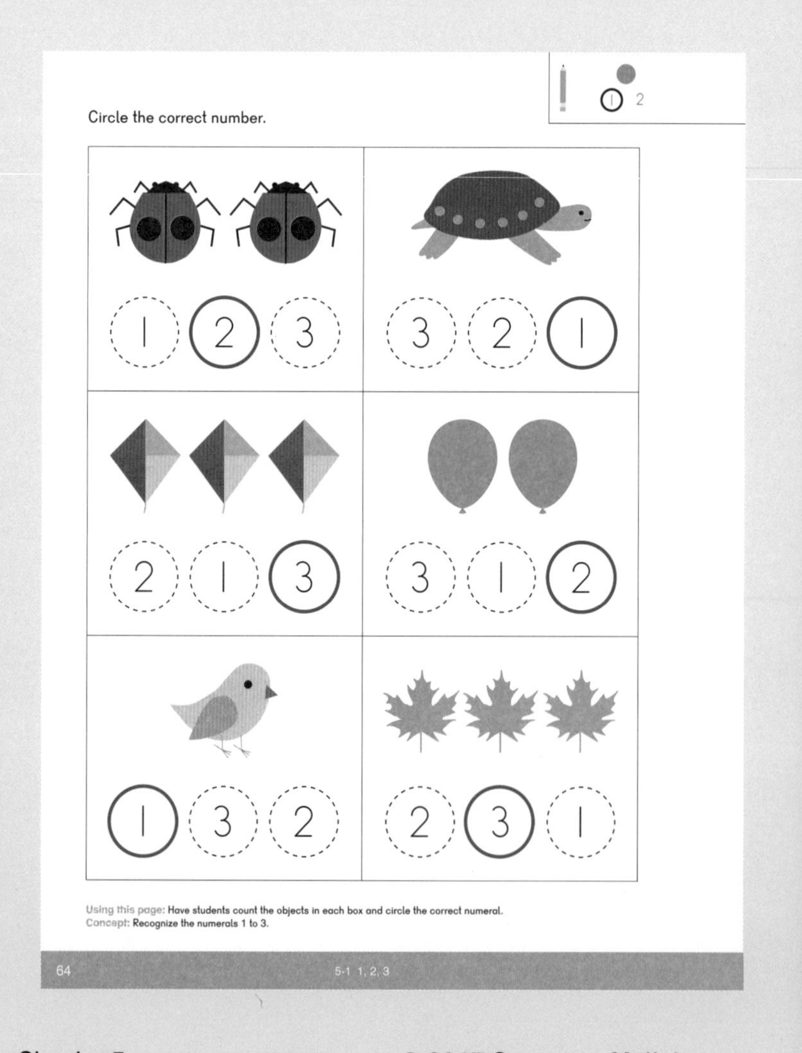
Circle the correct number.
1 2 3
3 2 1
2 1 3
3 1 2
1 3 2
2 3 1
Using this page: Have students count the objects in each box and circle the correct numeral.
Concept: Recognize the numerals 1 to 3.
64
5-1 1, 2, 3

Exercise 2 • pages 65 – 68

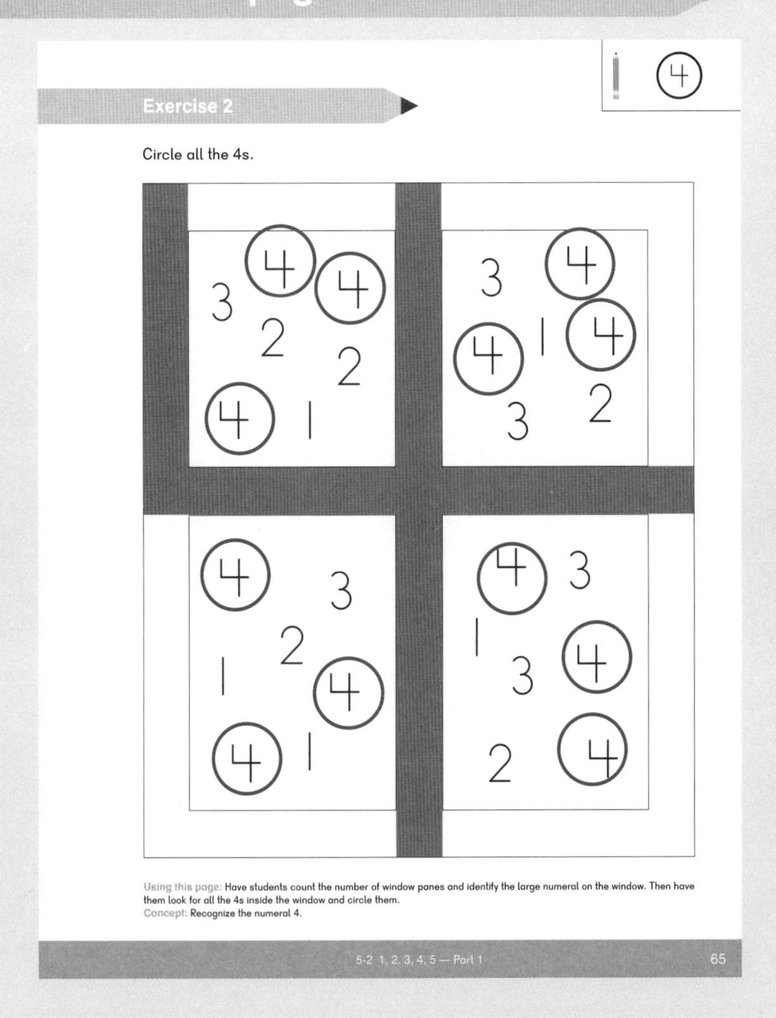

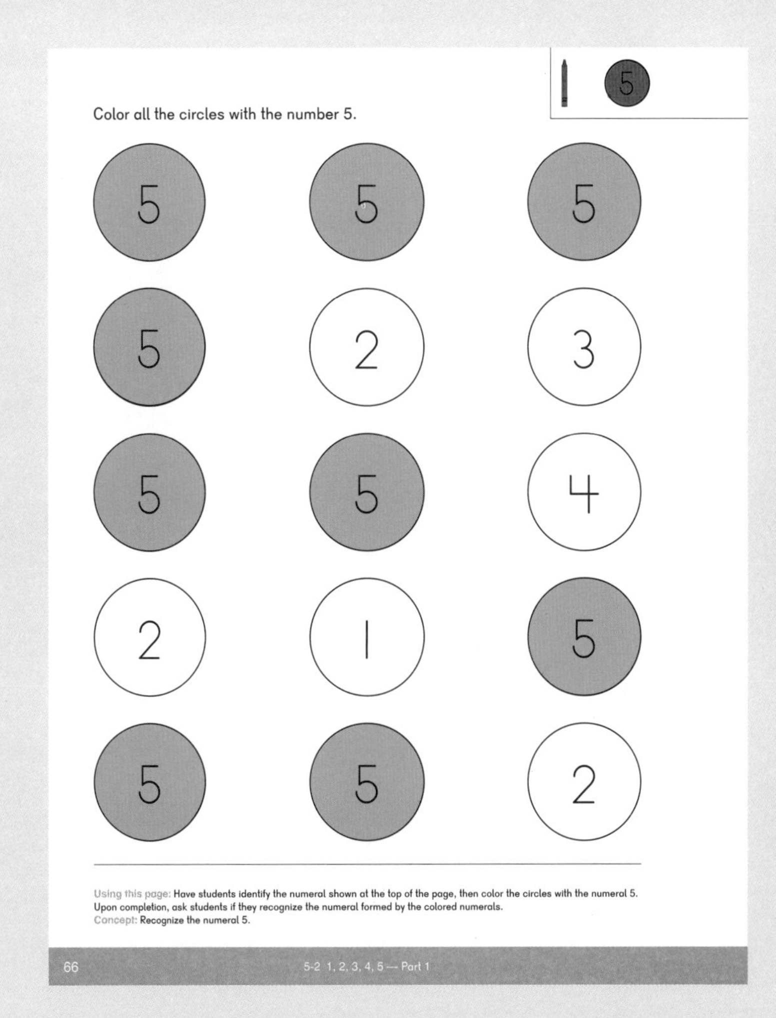

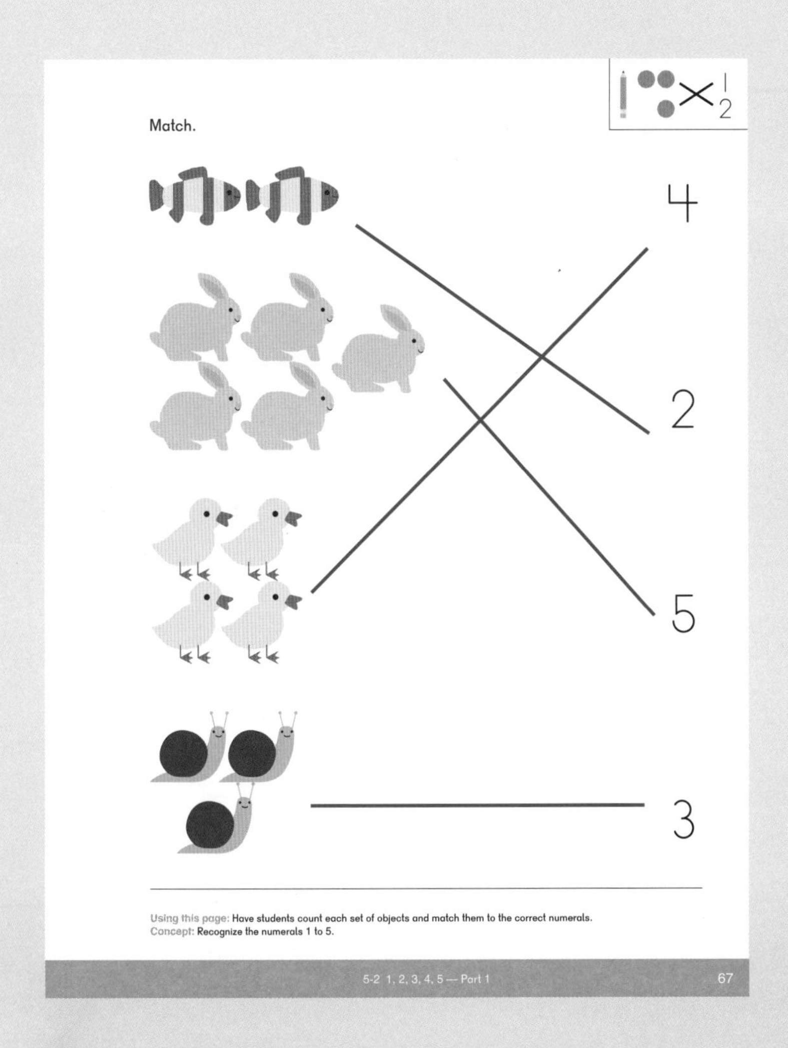

Exercise 3 • pages 69 – 70

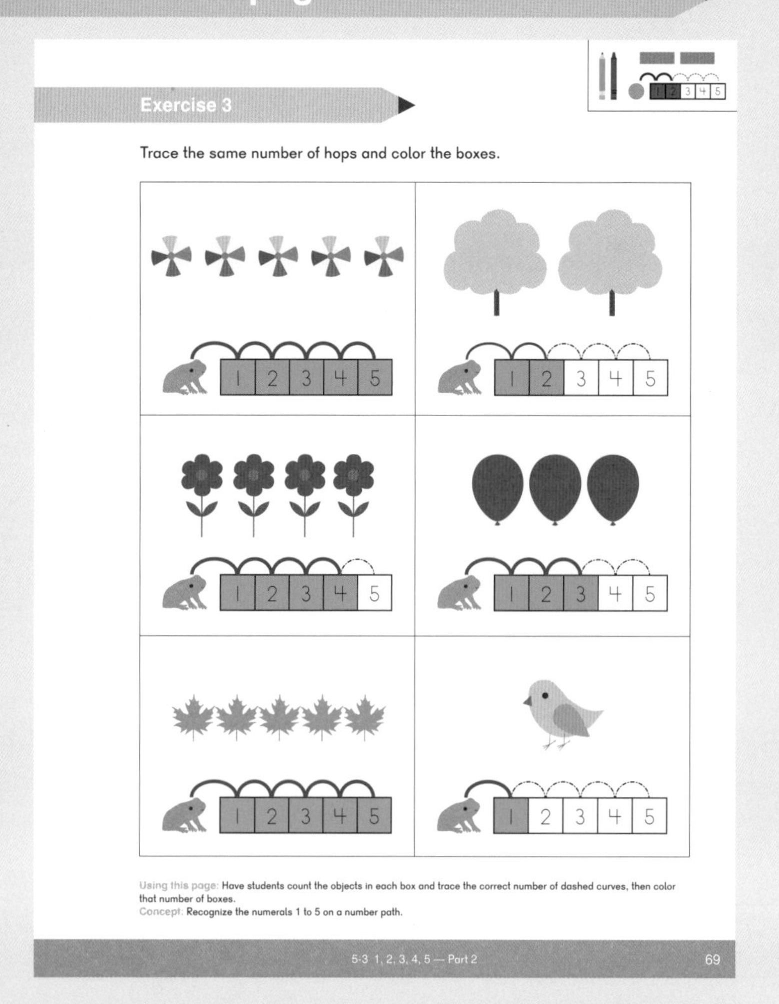

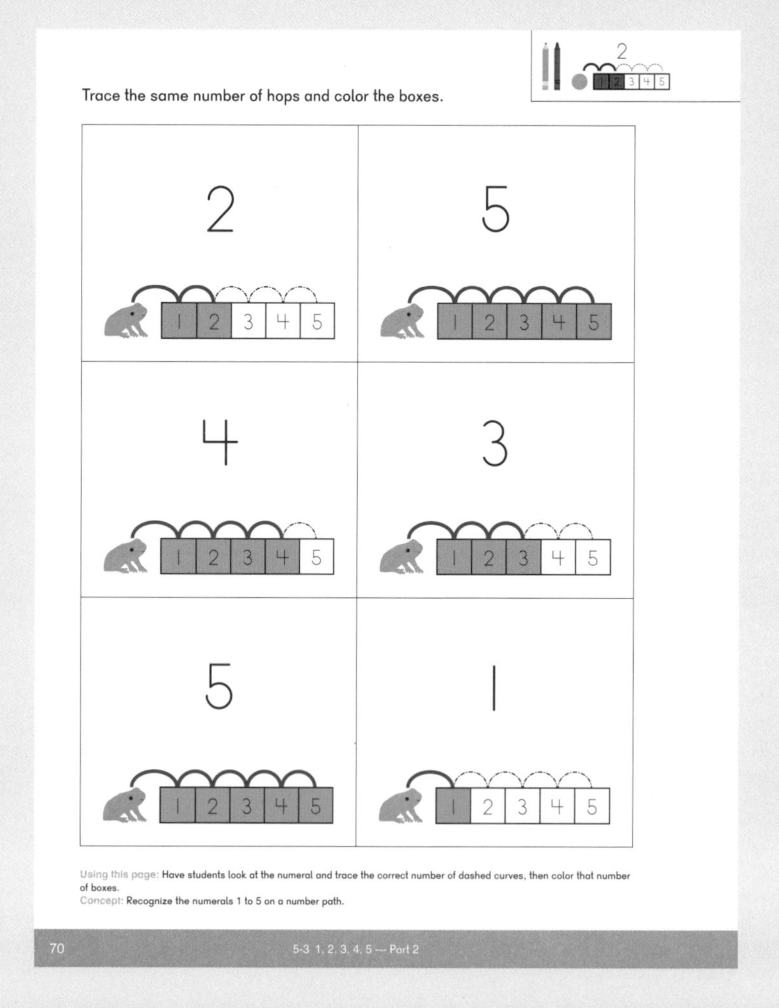

Exercise 4 • pages 71 – 72

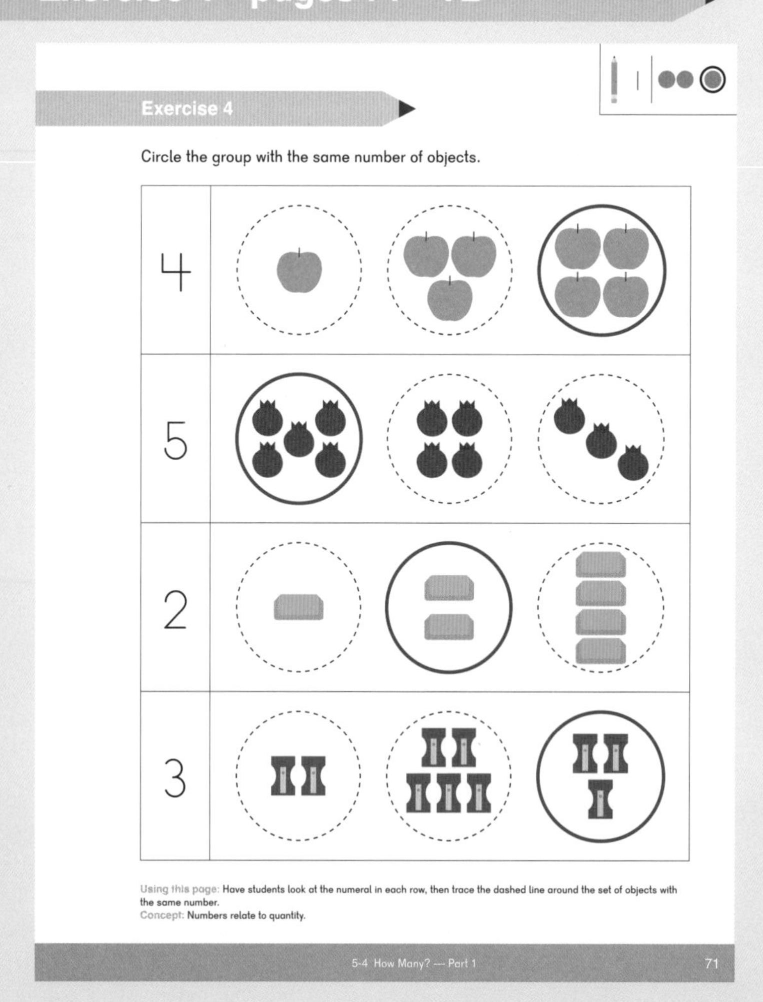

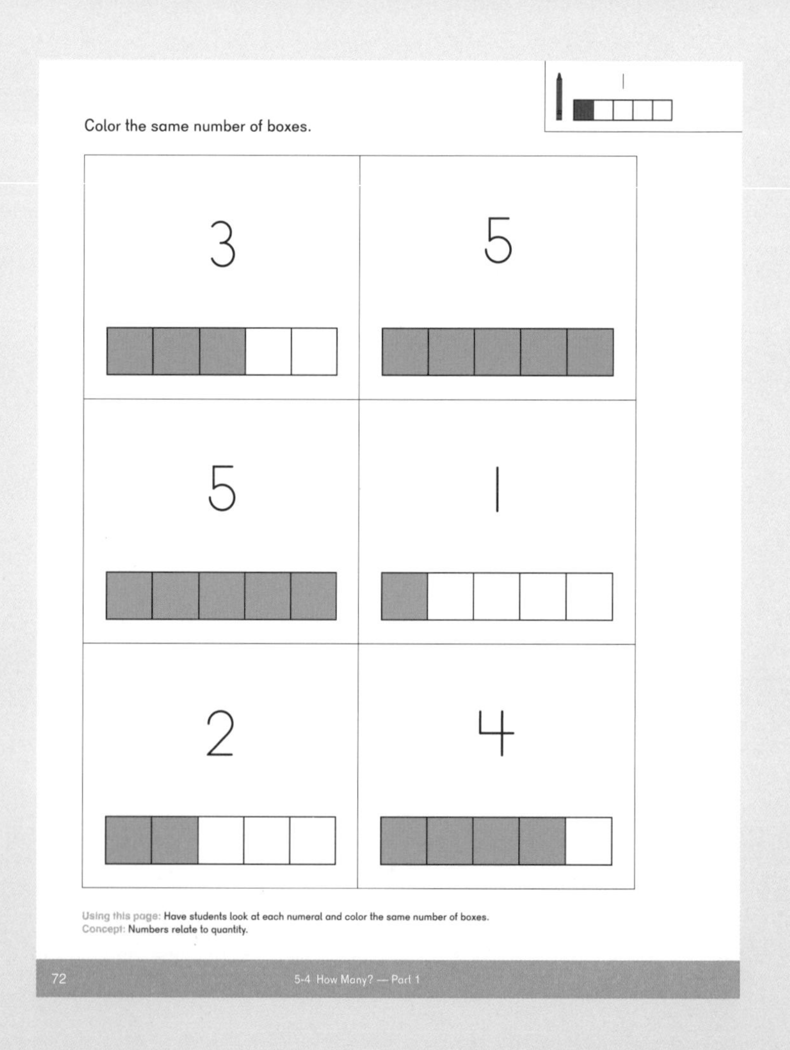

Exercise 5 • pages 73 – 74

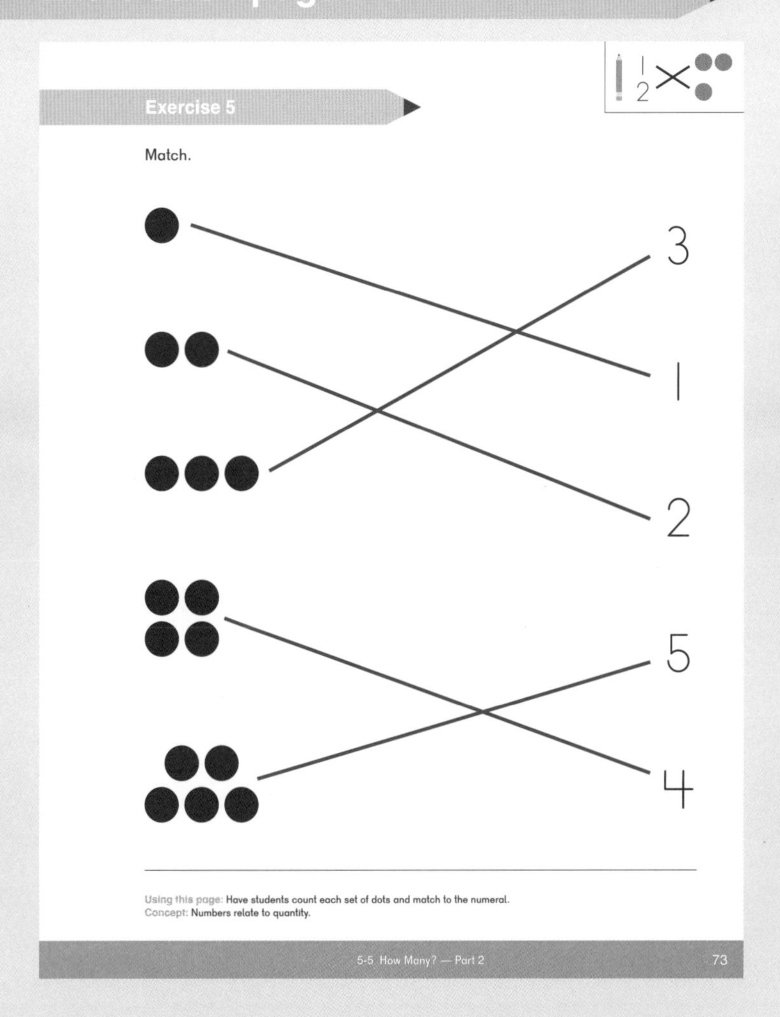

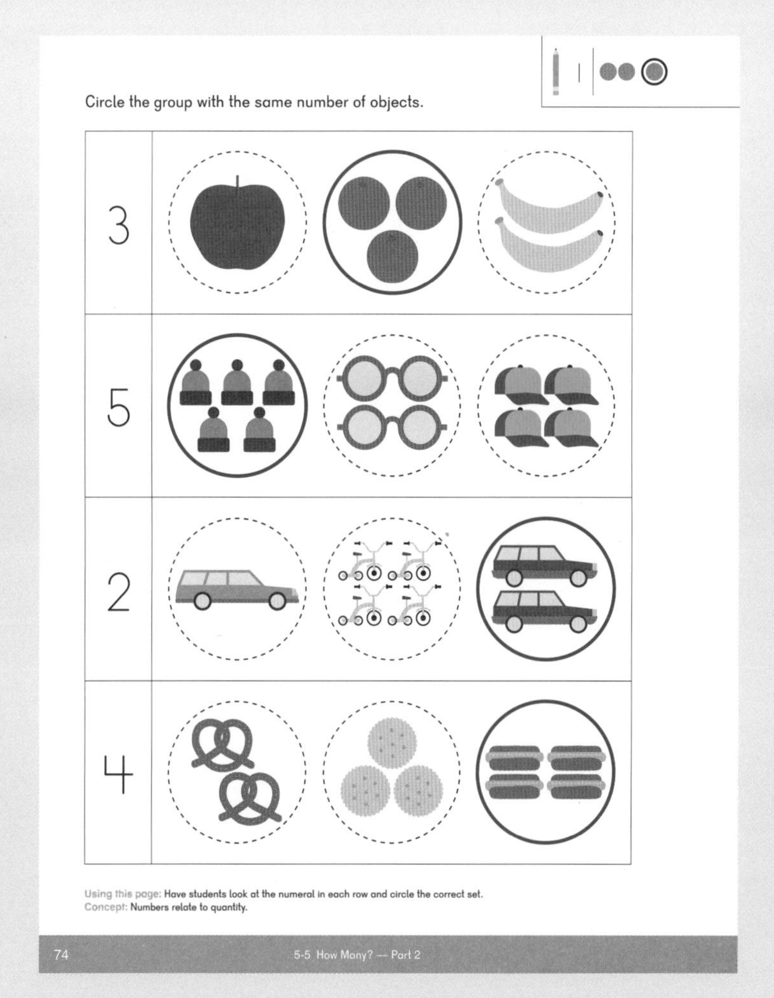

Exercise 6 • pages 75 – 76

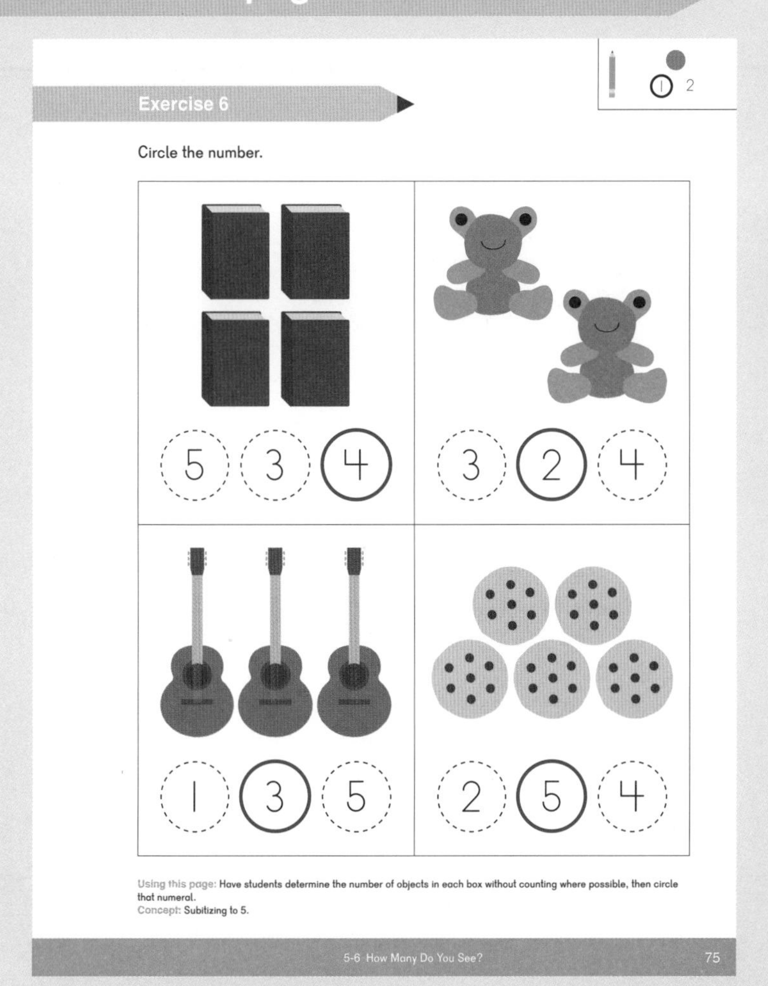

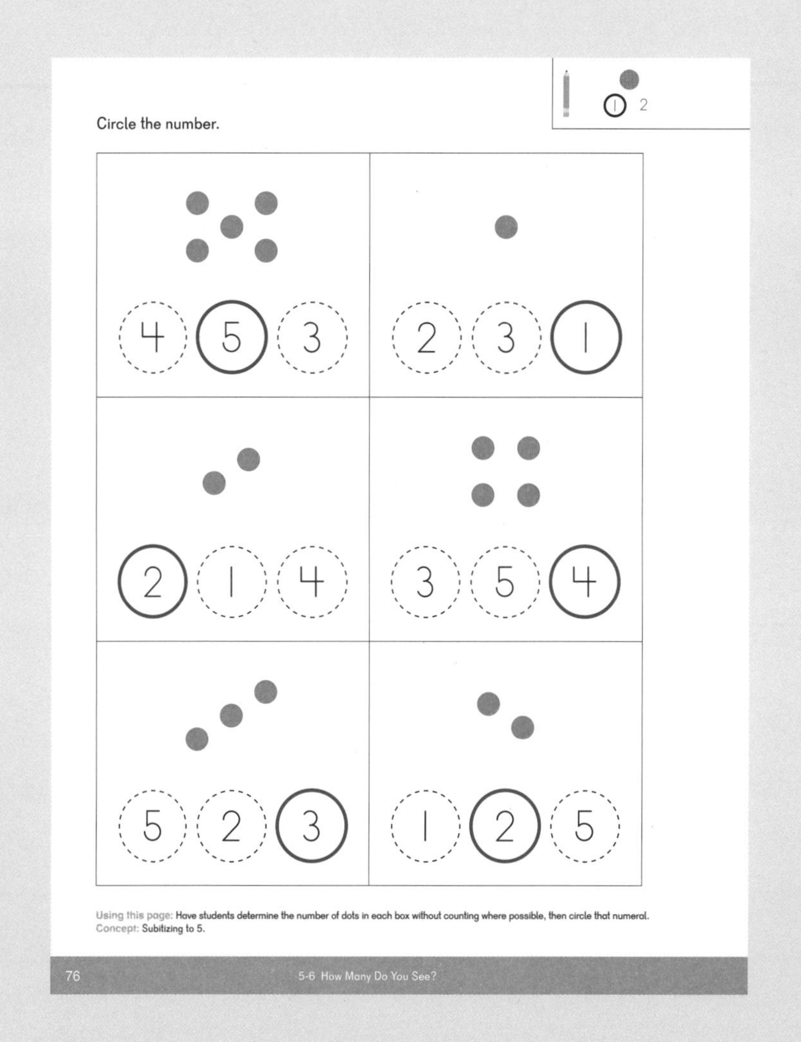

Exercise 7 • pages 77 – 78

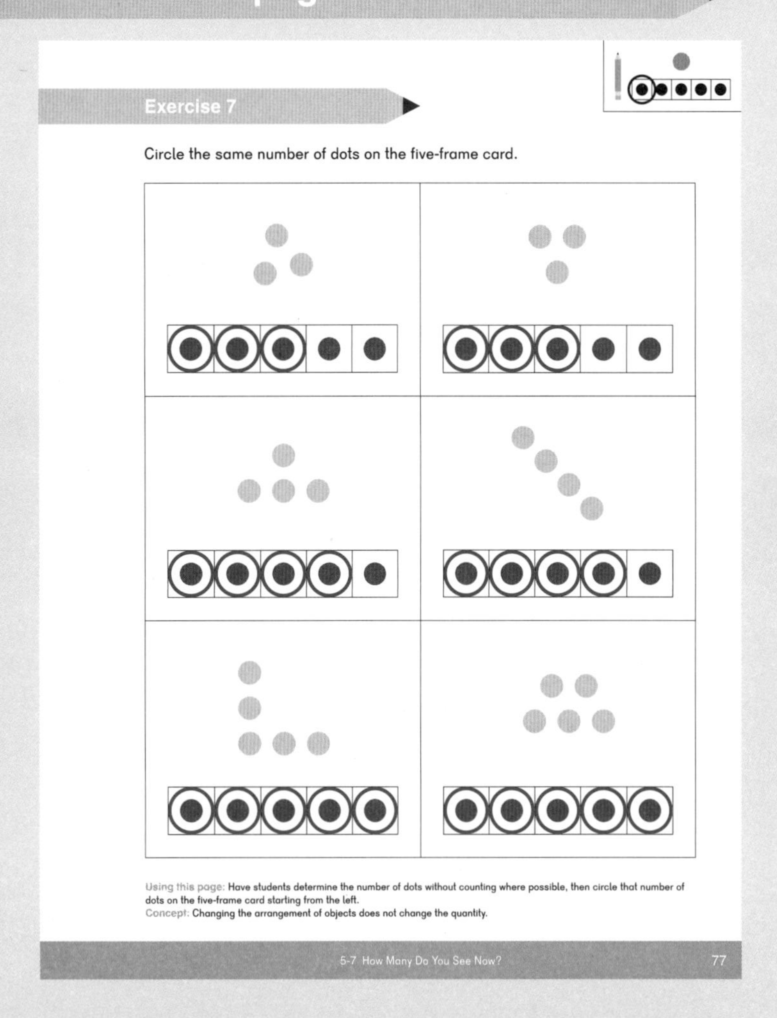

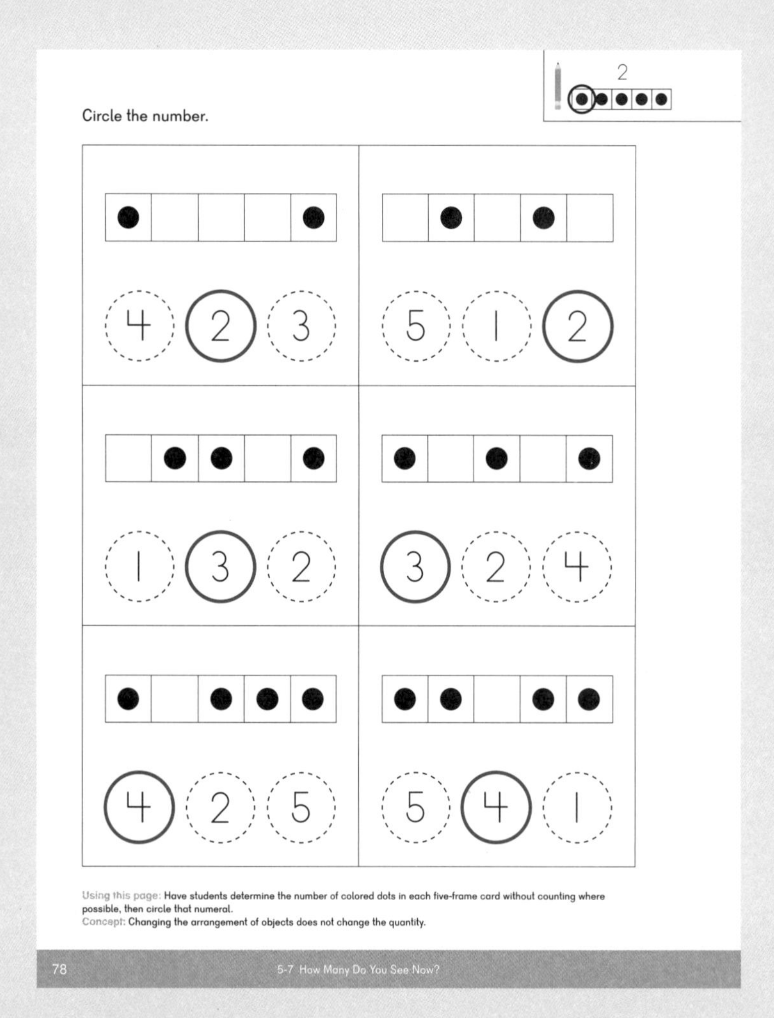

Exercise 8 • pages 79 – 80

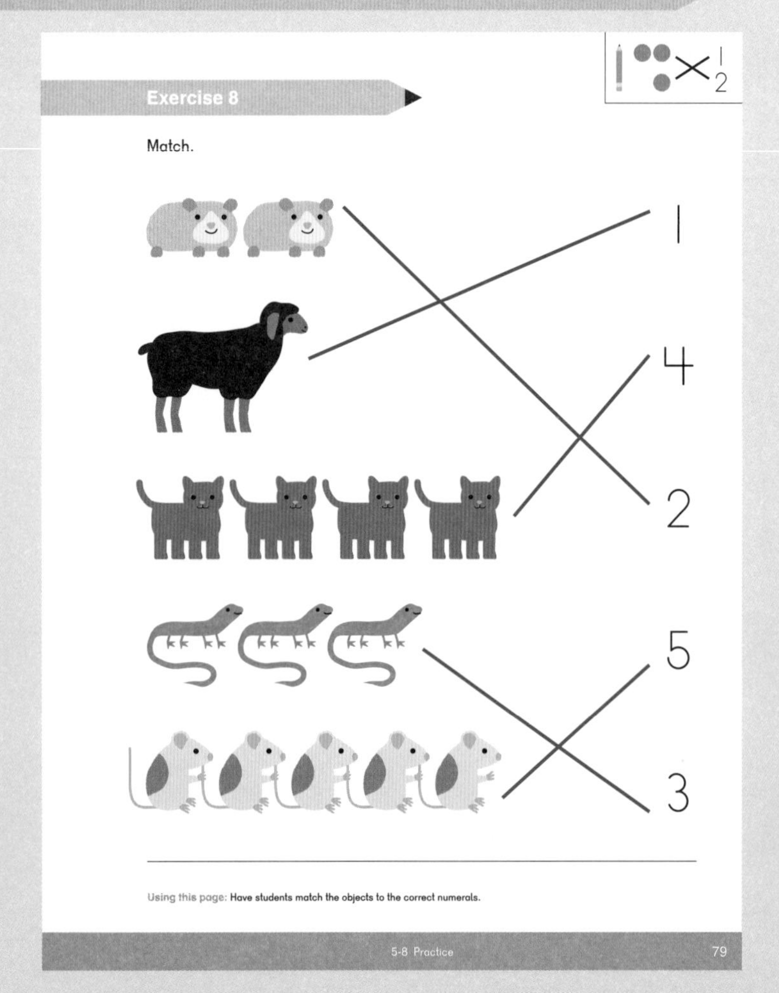

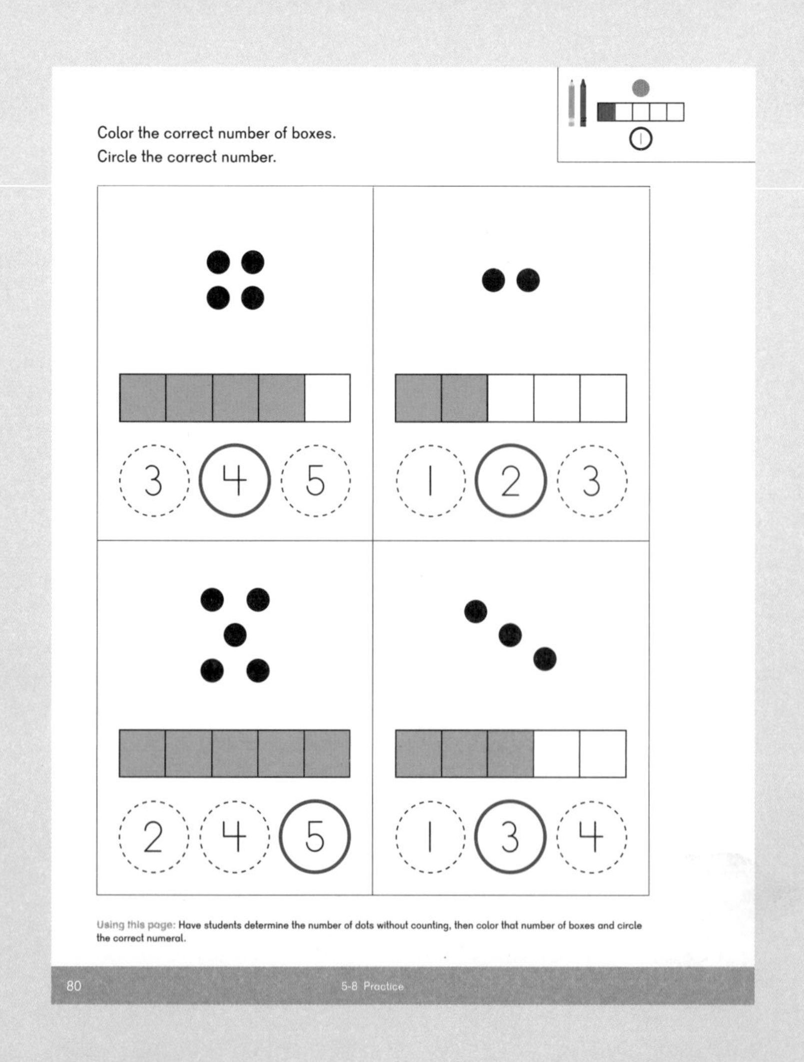

Suggested number of class periods: 13 – 14

Lesson		Page	Resources	Objectives
	Chapter Opener	p. 137	TB: p. 89	
1	0	p. 138	TB: p. 90 WB: p. 81	Understand that a set with no objects in it is a set of zero.
2	Count to 10 — Part 1	p. 140	TB: p. 93	Count to 10 by rote.
3	Count to 10 — Part 2	p. 142	TB: p. 94	Count to 10 by rote.
4	Count Back	p. 144	TB: p. 95	Count from 10 to 1 by rote.
5	Order Numbers	p. 146	TB: p. 97	Sequence numbers 0 to 10 and 10 to 0.
6	Count Up to 6 Objects	p. 148	TB: p. 98 WB: p. 83	Count up to 6 objects with one-to-one correspondence.
7	Count Up to 7 Objects	p. 150	TB: p. 100 WB: p. 85	Count up to 7 objects with one-to-one correspondence.
8	Count Up to 8 Objects	p. 152	TB: p. 102 WB: p. 87	Count up to 8 objects with one-to-one correspondence.
9	Count Up to 9 Objects	p. 154	TB: p. 103 WB: p. 89	Count up to 9 objects with one-to-one correspondence.
10	Count Up to 10 Objects — Part 1	p. 156	TB: p. 105 WB: p. 91	Count up to 10 objects with one-to-one correspondence.
11	Count Up to 10 Objects — Part 2	p. 158	TB: p. 107 WB: p. 93	Count up to 10 objects with one-to-one correspondence.
12	How Many?	p. 160	TB: p. 108 WB: p. 95	Count up to 10 objects with cardinality.
13	Practice	p. 162	TB: p. 110 WB: p. 97	Practice concepts introduced in this chapter.
	Workbook Solutions	p. 164		

Chapter Vocabulary

- Zero
- Six
- Seven
- Eight
- Nine
- Ten
- Ten-frame card
- Week
- Sunday
- Monday
- Tuesday
- Wednesday
- Thursday
- Friday
- Saturday
- More than

In **Chapter 4: Numbers to 5 — Part 1** and **Chapter 5: Numbers to 5 — Part 2**, students learned to count by rote, count with one-to-one correspondence, count with cardinality, and identify numerals 1 to 5. In this and the next chapter, students build on that prior knowledge, first learning about 0 and then learning about numbers 6 to 10. In **Chapter 7: Numbers to 10 — Part 2**, students will learn to recognize the numerals 6 to 10.

In **Dimensions Math® Pre-K** and **Kindergarten**, emphasis is placed on the anchor numbers, also known as benchmark numbers, 5 and 10. These numbers are focused on because they help students visualize quantities, subitize, and gain mental math skills. For example, the ten-frame card below helps students recognize a representation of 6 and understand that 6 is 1 more than 5.

●	●	●	●	●
●				

Ten-frames showing numbers 0 through 10 will be used extensively in the **Dimensions Math®** series. As illustrated above, ten-frames show two rows of five.

Key Points

Many students may already be familiar with the words zero, six, seven, eight, nine, and ten. Even so, they are included as vocabulary terms in this chapter because of the importance of associating the word with the quantity, which goes beyond rote recitation of these words when counting.

The days of the week are introduced in **Lesson 7: Count Up to 7 Objects.** Continue to practice the days of the week song throughout the year and emphasize that there are seven days in a week.

---

[1]Menken, A., Ashman, H., & Disney, W. (1990). *The Little Mermaid.* Milwaukee: Hal Leonard.
Goldfish® is a registered trademark of Pepperidge Farm.
Cracker Jack® is a registered trademark of Frito-Lay North America, Inc.

Materials

- Small counters
- Pair of gloves, or pictures of gloves
- Pair of toe socks or pictures of toe socks
- Large picture of an octopus
- Music: "Under the Sea" from "The Little Mermaid[1]"
- Bags, bowls, and baskets
- Linking cubes
- Magnets, small magnetic and non-magnetic objects
- Work mats
- Yardsticks
- String
- Metal paper clips
- Paper fish
- Plastic bottles
- Cardboard
- Hot glue
- Outer space decorations such as stickers, glow-in-the-dark stars, and aluminum foil
- Pictures of the solar system
- Small spheres
- Keyboard
- Egg cartons, a set with 6 divots and a set with 10 divots for each student
- Teddy bear counters
- Bell peppers
- Real or play vegetables
- Hula hoops or painter's tape
- Paper cups
- Large ten-frames showing numbers 0 to 10
- Stickers
- Baking tins
- Play cookies
- T-ball bat, ball, bases
- Paper shapes: circles, rectangles, and squares
- Unsharpened pencil
- Clear plastic cups
- Dress-up clothing items including "grown-up" office clothes and buckle shoes or belts
- Pretend bowling pins and ball or toy bowling set
- Beach ball
- Cuisenaire rods
- Large yellow pom-poms
- Googly eyes
- Rocks to be painted

Note: Materials for Activities will be listed in detail in each lesson.

Blackline Masters

- Ten-frame Cards 0 to 10
- Blank Ten-frame
- Keyboard Template
- Vegetable Cards
- Octopus Template
- Cuisenaire Rods

Storybooks

- *Ten in the Bed* by Penny Dale
- *The Very Hungry Caterpillar* by Eric Carle
- *Ten Eggs in a Nest* by Marilyn Sadler

Optional Snacks

- Biscuits
- Windmill cookies
- Goldfish® crackers
- Rice cereal treats cut into moons or stars
- Finger sandwiches
- Carrot sticks
- Cherry tomatoes
- Green grapes
- Round crackers
- Pretzel sticks
- Cracker Jack®
- Cream cheese
- Apple slices

Letters Home

- Chapter 6 Letter

Notes

Lesson Materials

- Ten-frame Card (BLM) showing 10
- Pair of gloves or pictures of gloves
- Pair of toe socks, or pictures of toe socks
- Ten-divot egg carton

Explore

Show a pair of gloves clipped together, a ten-divot egg carton, a Ten-frame Card (BLM) showing 10, and a pair of toe socks clipped together. Ask students what is alike about everything you are showing. Have them count with you the fingers in the gloves, the divots in the egg carton, the colored dots on the ten-frame card, and the toes in the toe socks.

Learn

Have students look at page 89 and identify the objects on the page. Ask questions such as:

- Why do you think there are two eyes and only one nose?
- How many leaves on the shamrock?
- How many squares?
- What things on this page have the same number as there are eyes?
- Can you find and point to the number on the path that is the same as the number of leaves on the shamrock?
- Can you find pictures in the classroom that have the same number as the egg carton we just counted?

Chapter 6

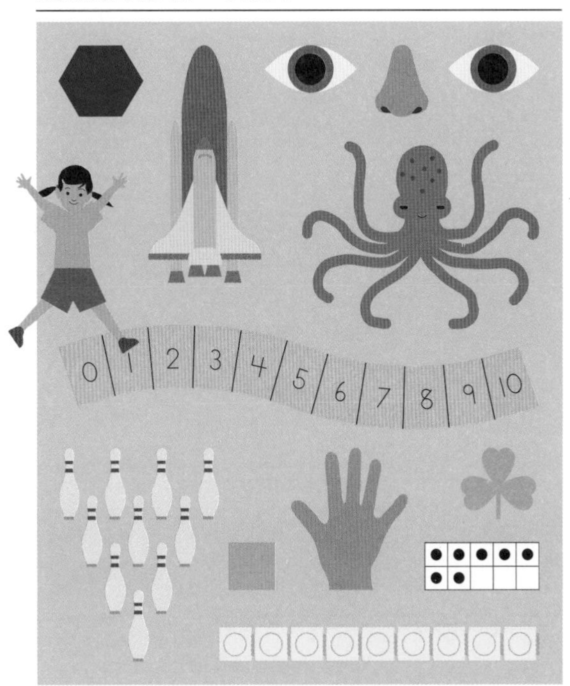

89

Extend Play

Octopus Adventure: Play the song "Under the Sea" from the movie *The Little Mermaid*[1] and talk about what types of creatures live in the ocean. Show a large picture of an octopus. Talk about octopus arms and count the arms together. Divide students into groups of four. Have each group act as if their eight arms are the arms of an octopus and move around the classroom.

Materials: Large picture of an octopus, recording of "Under the Sea" from "The Little Mermaid[1]" if possible

[1]Menken, A., Ashman, H., & Disney, W. (1990). *The Little Mermaid.* Milwaukee: Hal Leonard.

Objective

- Understand that a set with no objects in it is a set of zero.

Lesson Materials

- 5 toys
- Basket
- Bags containing 5 linking cubes, 1 bag per student
- Optional snack: Biscuits

Explore

Show students five toys on the floor. Ask how many toys are on the floor. Tell them that you want to put the toys away in a basket, and that you want to know how many toys are still on the floor each time you put one in the basket. Have them count back from five with you as you put the toys in the basket. When you are done, ask them how many toys are on the floor. Tell them that the math word that means "none" is "zero." Write the word "zero" and a "0" for students to see.

Give each student a bag of linking cubes. Have them pour the cubes on the floor. Tell them that they will be pretending that the cubes are doggie biscuits, and they will be putting the pretend doggie biscuits away in the bag. Have them say, "Four left on the floor, three left on the floor," etc. with you as they place the cubes back in the bag.

Learn

Have students look at page 90. Introduce Spot and have students say how many doggie biscuits Spot has in his bowl.

Have students look at page 91. Ask them what they think is happening on the page. Then read Mei's comments.

Have students look at page 92 and read them Sofia's comment at the top of the page. Ask students to describe the appearance of 0 in their own words. Read Sofia's direction and have students complete the task.

Whole Group Play

Drop 'Em!: Give each student five objects that don't roll easily, such as building blocks. Lay the hula hoops on the floor, or use painter's tape to make enclosed areas. Count from 5 to 0 with students as they drop the objects into the hula hoops.

Materials: 5 non-rolling objects per student, several hula hoops, painter's tape

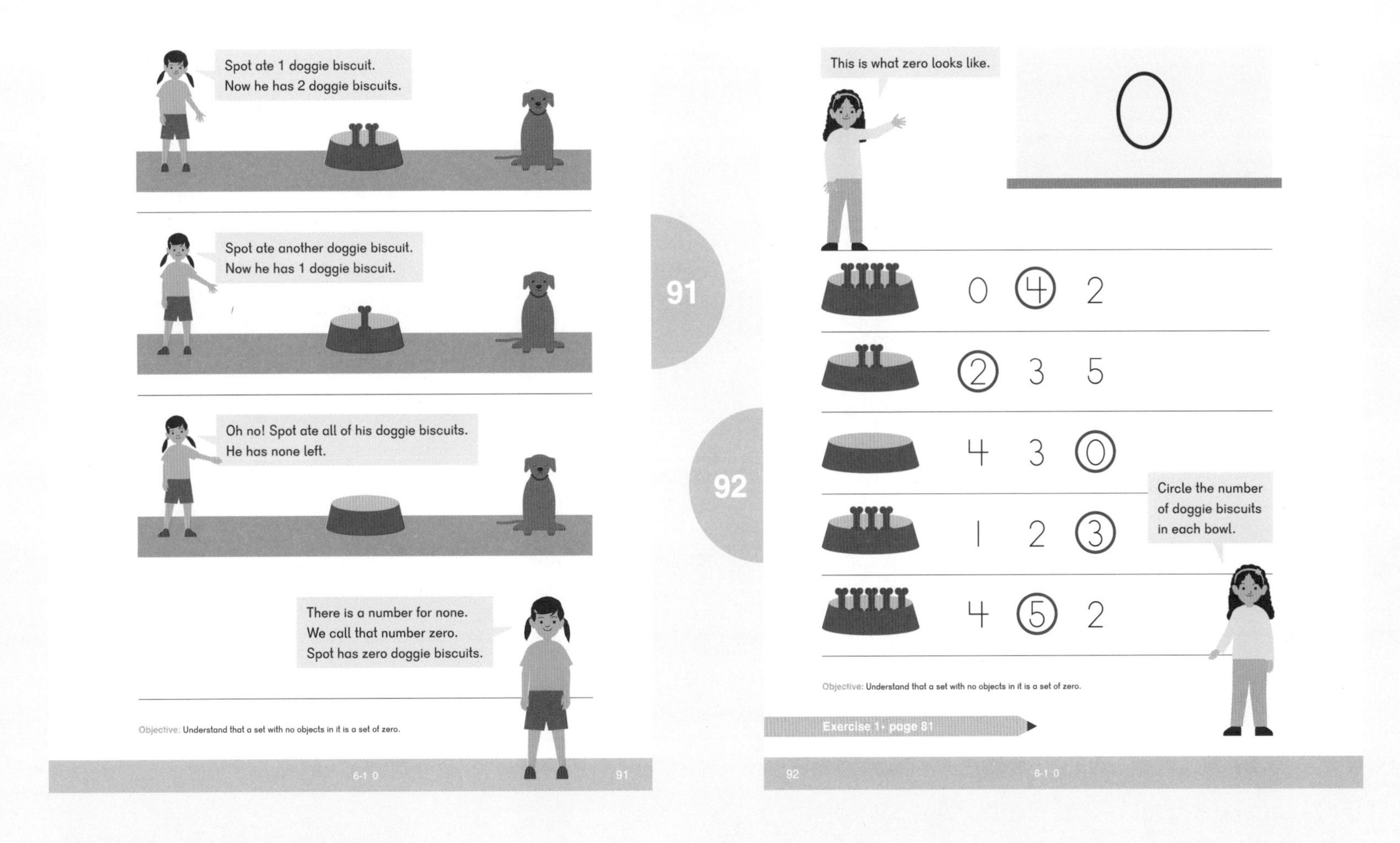

Small Group Center Play

Sort: Provide bowls labeled 0 to 5. In a container, have one red crayon, two blue crayons, etc., to five of a type of crayon. Have students sort the crayons into the bowls with the corresponding number. For example, two blue crayons go into the 2 bowl. After sorting, have students go over how many of each type of crayon were in the container. Ask about a color of crayon that was not there, encouraging them to use the word "zero." This activity can be done with any type of object.

Hungry Spot: Have students count out 3 counters and put them in a bowl. Students pretend to be Spot and move the counters one at a time from the bowl. They can bark and look sad when all of the biscuits are gone.

Draw Spot: Have students draw 2 pictures of Spot: One portraying him when he still has doggie biscuits, and another after he has eaten all of them.

Exercise 1 • page 81

Extend Learn

All Gone: Have students tell stories to each other in which the quantity left at the end of the story is zero.

Objective

- Count to ten by rote.

Lesson Materials

- How Much is that Doggie in the Window? (VR)
- Optional snack: Windmill cookies

Explore

Have students clap their hands 10 times with you. Count the claps. See how many students can already count to 10. Exaggerate the numbers 6 to 10 while counting. Repeat with clapping hands on legs. Teach students the song on page 93, sung to the tune of "How Much is that Doggie in the Window" (VR). Sing several times using different voices and volumes.

Learn

Have the students look at page 93. Ask them what they remember about each friend. Point out the fingers. Sing the song again. Then rote count from 1 to 10 with the students several times, whispering, shouting, clapping, etc. while counting.

Lesson 2
Count to 10 — Part 1

One two three four fi-ive si-ix seven
Eight nine ten, I counted to ten.
One two three four fi-ive si-ix seven
Eight nine ten, I counted again.

93

Objective: Count to 10 by rote.

Whole Group Play

Move for 10: Lead students in several exercises, such as jumping jacks, rote counting to 10 each time.

Small Group Center Play

Sort: Provide containers holding up to 10 of seven different types of objects and bowls to sort into. The groups should contain four of one object, five of a different object, etc., up to 10. Have students sort the objects in more than one way, then count and say the number in each group.

My Name Is _____: Have students dress up like one of the friends and act like that character.

Art: Have students draw a picture of one of the friends showing his or her fingers.

No exercise for this lesson.

Extend Explore

10 Around the World: Teach (or have students or other teachers teach) students to count to 10 in a different language.

Objective

- Count to 10 by rote.

Lesson Materials

- Magnets
- Bags containing magnetic and non-magnetic small objects, 1 bag per pair of students
- Work mats, 1 per pair of students
- Yardsticks
- String
- Metal paper clips
- Optional snack: Goldfish® crackers

Explore

Give each pair of students a magnet, work mat, and bag of objects. Allow them to explore for several minutes. After allowing students to explore, ask them to discuss what they noticed about the magnets and the objects.

Have students stand in a circle and show them a fishing pole made by tying a string to a yard stick and tying a small magnet to the end of the string. Teach students the rhyme from the textbook.

As you teach them the rhyme, use the fishing pole to which you have attached a metal paper clip, to catch a paper fish. Show students your right pinkie when you read that word in the rhyme.

Learn

Have students look at page 94 and say what Mei is doing. Read Emma's comment aloud. Ask students if they've heard of people catching fish and letting them go for any reason other than because the fish bit their fingers. Read the rhyme again and have students act it out.

Whole Group Play

Count in Funny Voices: Have students stand in a circle. Tell students that they will be counting aloud to 10 using funny voices. Have one student say, "One." The student next to the starting student says, "Two," etc. Whisper the correct number to any student who is unsure. Assess which students have not yet mastered rote counting to 10.

Magic Number: Have students sit in a circle and count to 10. The first student says "One," the next student says, "Two," etc. The student who says, "Ten" stands up and then the counting starts again until everyone is standing up. Standing students continue to participate in the counting.

Small Group Center Play

Sort: Provide containers holding up to 10 of several different types of objects and bowls to sort into. Have students sort the objects in more than one way.

Go Fish: Give each pair of students a fishing pole and 10 paper fish. Have students take turns "fishing" as they say the rhyme together.

Hand Print Fish: Have students dip their hand in paint and transfer the paint to a piece of paper. After the paint has dried, have students glue googly eyes onto the fish and draw an underwater scene.

No exercise for this lesson.

Extend Learn

Magnetic or Not?: Have students guess which objects in the classroom are magnetic, then have them use a magnet to see if they are correct.

Materials: Magnets

Objective

- Count from 10 to 1 by rote.

Lesson Materials

- Bags containing 10 small counters, 1 per student
- Blank Ten-frames (BLM), 1 per student
- *Ten in the Bed* by Penny Dale
- Optional snack: Rice cereal/marshmallow treats cut into moons or stars

Explore

Give each student a bag of counters and a Blank Ten-frame (BLM). Have them place their counters on their ten-frames, starting in the upper left corner of the card, filling in the top row, then starting the bottom row from the left.

Either read the book *Ten in the Bed* by Penny Dale or teach the song “Ten in the Bed” to the students (VR). Choose 10 students to lie next to each other on the floor and get ready to act out the story. Have other students model the story with their counters on their ten-frame cards.

As you read or sing, each time “one rolls out,” a student should roll away from the others. Students not acting will remove one counter at a time from their ten-frame cards.

After finishing the first read-through or song, choose other students to act out the story. This time, each time one rolls out, ask, “How many are still left in the bed?”

Collect the counters and ten-frames.

Tell students to hold up all 10 fingers and count them with you, starting with their left pinkies.

Read or sing the story again, having students put down one finger at a time, starting with their right pinkies.

Learn

Have students look at page 95 and count the students by putting a finger on each. Then have them identify the five friends.

Have students look at page 96 and discuss the page. Have them look at the numbers written near Sofia and identify as many of them as they can.

Whole Group Play

Blast Off!: Have students stand, clasp their hands over their heads, and count back from 10 to 1 as they lower their bodies. When they get to, “One,” have them jump in the air as they say, “Blast off!”

Outer Space Sensory Tables: Use black sand or gravel, glow-in-the-dark stars, and different size spheres, such as marbles, ping pong balls, small bouncy balls, foam balls, etc. to create outer space sensory tables. Note: Black beans will work instead of black sand or gravel, if necessary. Keep **Outer Space Sensory Tables** in place for use in **Lesson 5: Order Numbers.**

Have students use their hands to explore in the sand, gravel, or beans until they find a star or a sphere. Before removing the object from the sand, gravel, or beans, have them describe the feel of the object, then identify it as either a star or a ball by its feel.

As each object is described, have the student describing it remove it from the sand, gravel, or beans.

No exercise for this lesson.

Small Group Center Play

Sort: Different sizes of small spheres by size.

Fly Me to the Moon: Ask students what they know about space, astronauts, and rocket ships. Play **Blast Off** again.

Build Rocket Ships: Use plastic bottles covered in construction paper for the ship and hot-glue cardboard to each bottle for the wings. Have students decorate their rocket ships with stickers, crayons, etc. After rocket ships are built, take students outside and have them spread out with their rocket ships. Count down from 10 to 1 chorally, then have students launch their rockets by raising them off the ground and tossing them straight up in the air.

Extend Play

Space Story: Have students record a story about an astronaut.

Materials: Recording device

Lesson 5 Order Numbers

Objective

- Sequence numbers 0 to 10 and 10 to 0.

Lesson Materials

- Keyboard for teacher, if available, otherwise Keyboard Template (BLM)
- Keyboard Template (BLM) for each student
- Magic Thumb (VR)
- Optional snack: Finger sandwiches

Explore

Play or sing 10 musical notes from low to high. For example, if a keyboard is available, start on middle C and end on E an octave higher. If singing the notes without a keyboard, demonstrate using your fingers on the Keyboard Template (BLM) starting with your left pinkie.

Play notes again, singing a number 1 to 10 for each note.

Give each student a keyboard template. Have them place their left pinkies on C, left ring finger on D, etc., until their right pinkie is on E. Have them pretend to play a note on their templates each time you sing a number. Have them repeat, singing the numbers with you. Repeat. The purpose of this is to link each of their fingers to a number.

Note: There are many keyboard apps available for your use.

Teach students how to play **Say the Number and Win.** Two players are trying to say a target number by counting on or counting back. You call out a number between 3 and 10. The first player says, “One,” the other player says, “Two,” etc., until one of them says the number you called. That player is the winner.

Repeat with different numbers.

Change the game by having the first player say, “Ten,” the second player say, “Nine,” etc., until one of them says the number you called.

Play Magic Thumb (VR).

Learn

Have students look at page 97. Tell them that Alex and Dion are playing **Say the Number and Win.**

Read Alex’s and Dion’s comments aloud. Ask them who will say, “Seven” if Alex starts. Repeat with other numbers.

Whole Group Play

Say What?: Tell students that you will be counting from 0 to 10 or from 10 to 0. Say, "Zero, one, two, four, five." Wait for students to tell you that you made a mistake and that you skipped the number three. Repeat several times, skipping different numbers each time.

Small Group Center Play

Sort: Provide containers holding up to 10 of several different types of objects and bowls to sort into. Have students sort the objects in more than one way.

Symphony of Numbers: Have students pretend to play musical notes on their keyboard templates while singing the numbers in order from 1 to 10 and 10 to 1.

Painted Piano: Provide black and white paint. Have students paint keyboard keys by looking at their Keyboard Templates (BLM) and copying them.

No exercise for this lesson.

Extend Explore

Say the Number and Win (With a Twist): Have students predict who will say which number. Students can use any strategy to make their predictions. Some students may count forward quickly. Others may recognize a pattern of odd and even numbers.

Buried Treasure Hunt: Bury the objects in sand or gravel at the sensory table. Have students remove 10 different objects, knowing they are different because of how they feel. Then have students bury the objects again, counting back from 10 as they do so.

Materials: 20 small different objects, sensory table, sand or gravel

Lesson 6 Count Up to 6 Objects

Objective

- Count up to 6 objects with one-to-one correspondence.

Lesson Materials

- Music
- Blank Ten-frames (BLM), 1 per student
- Small containers of teddy bear or other counters, 10 in each, 1 container per student
- Real or play vegetables
- Containers to hold vegetables
- Optional snack: Carrot sticks

Explore

Tell students that you are going to play music. While the music is playing they are to walk in a circle. When the music stops, they are to get into groups of five as quickly as possible and sit down together. If your number of students is not a multiple of five, have the "extra" students sit together and say how many are in their group.

Ask students what number comes after five. Tell them that you are going to play music again, but when the music stops they are to get into groups of six and sit down in their groups. Repeat previous activity if your number of students is not a multiple of six.

Give each student a Blank Ten-Frame (BLM) and a container of counters. Show students how to place the ten-frame horizontally in front of them. Count out six counters and arrange them on your ten-frame, top row first left to right, then left corner on bottom row. Have students do the same. Ask them how many counters they are showing on their ten-frames.

Collect the ten-frames and counters. Place containers of real or play vegetables near them. Have each student count out six vegetables, then replace them in the containers, one at a time, counting back from six.

Learn

Have students look at page 98 and discuss Emma's comment. Have them identify the number of dots on each ten-frame. Encourage them to do so without counting.

Have students look at page 99 and identify the vegetables in each basket. Read Emma's direction. Have students count the eggplants and circle the basket. Then have them complete the task.

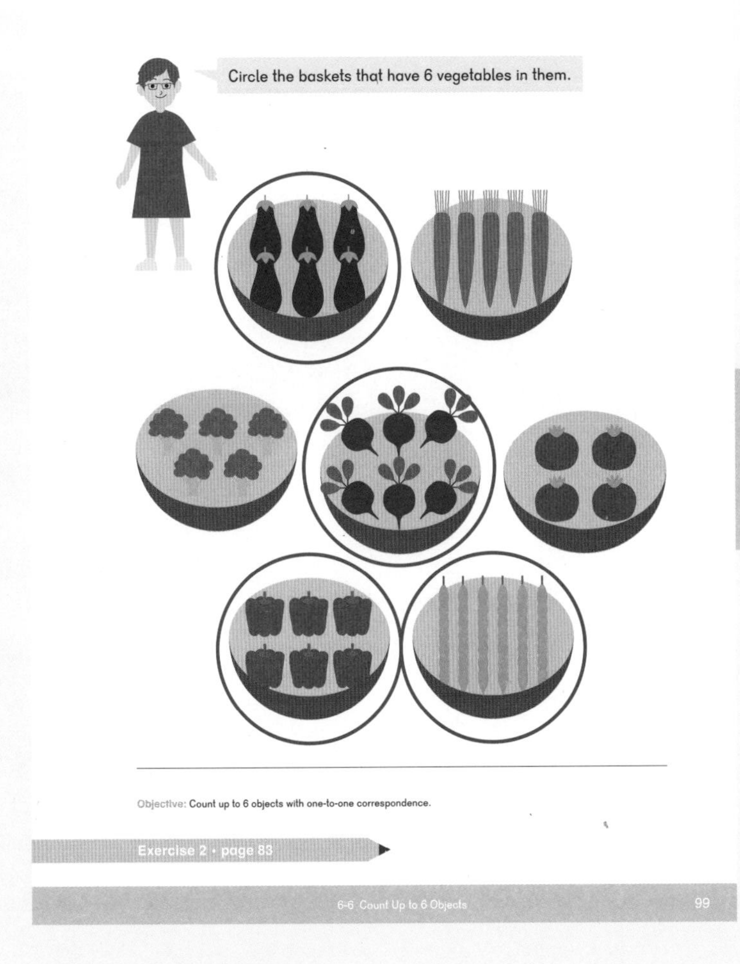

Whole Group Play

Simon Says: Play **Simon Says** with movements up to six.

Small Group Center Play

Sort: Provide containers holding up to 10 play or real vegetables of several different types and bowls to sort into. Have students sort the objects in more than one way.

Farm Stand: Set up a farm stand with cut-out Vegetable Cards (BLM). Have each student pretend to buy six of one of the types of vegetables by counting out that many paper cutouts of the vegetable. Give each student a blank ten-frame card. Have students color their vegetables the common color for that vegetable and arrange their vegetables on their ten-frame cards.

Bell Pepper Art: Cut the end off of several bell peppers, leaving the stem intact as a handle. Have students dip the pepper into paint, then stamp the pepper on the paper.

Counting: Provide small objects and 10-divot egg cartons. Have students count out 6 objects and place them in the egg carton as if placing them on a ten-frame, counting each as it is placed. Then have them remove the objects, counting back from 6.

Exercise 2 • page 83

Extend Play

Six Little Snails: Teach students the nursery rhyme.

Six little snails
Lived in a tree,
Johnny threw a big stone,
Down came three!

Ask students to discuss why they think only three snails fell out of the tree. What if Johnny used a different size stone? Have them act out the rhyme.

Objective

- Count up to 7 objects with one-to-one correspondence.

Lesson Materials

- Blank Ten-frames (BLM), 1 per student
- Bags containing 10 small counters, 1 bag per student
- Pop Goes the Weasel (VR)
- Calendar
- Optional snack: Fruity caterpillars: 6 green grapes and a cherry tomato for the head on skewers

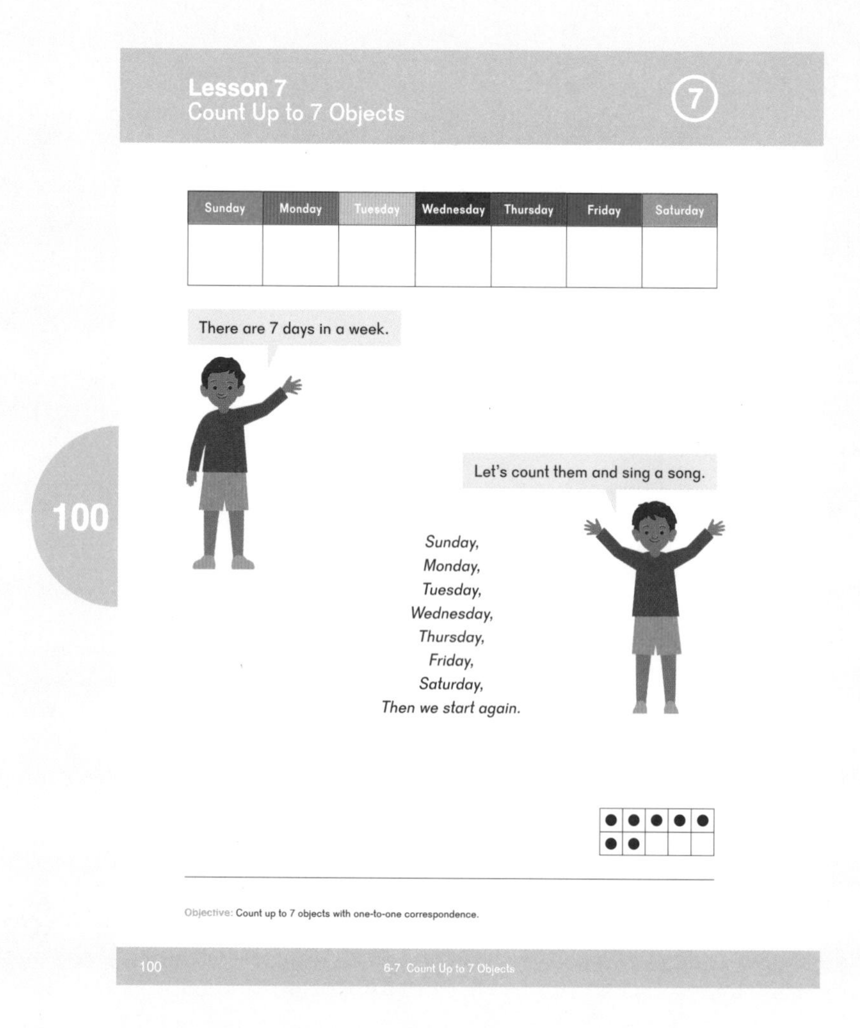

Explore

Give each student a Blank Ten-Frame (BLM) and a bag of counters. Have each student count out six counters and place them correctly on their ten-frames. Ask them, "What number comes after six?" Have them place one more counter on their ten-frames. Have them tell you how many counters they are showing. Collect the counters and ten-frames.

Tell students that they will be learning about days of the week today. Have them talk with a partner about what a day is. Suggested questions:

- How does your day start?
- During the day, does it seem like the sun moves in the sky?
- What do you do after the sun goes down?

Teach students the "Days of the Week Song" on textbook page 100, sung to the tune of "Pop Goes the Weasel" (VR). Explain that the days of the week are named Sunday, Monday, Tuesday, Wednesday, Thursday, Friday, and Saturday, and are always in the same order. Sing the song with them several times. Point to a calendar and sing the song again as you point to the names of the days. Count with students how many days are in a week.

Learn

Have students look at page 100. Read Alex's comments. Have students put a finger on each day of the week as the name of the day is sung as they sing the song with you again.

Have students look at page 101 and read Alex's direction aloud. Have them identify the objects on the page and try to say how many are in each set without counting. Ask them what they did to make it easier to count the objects. Then have them complete the task.

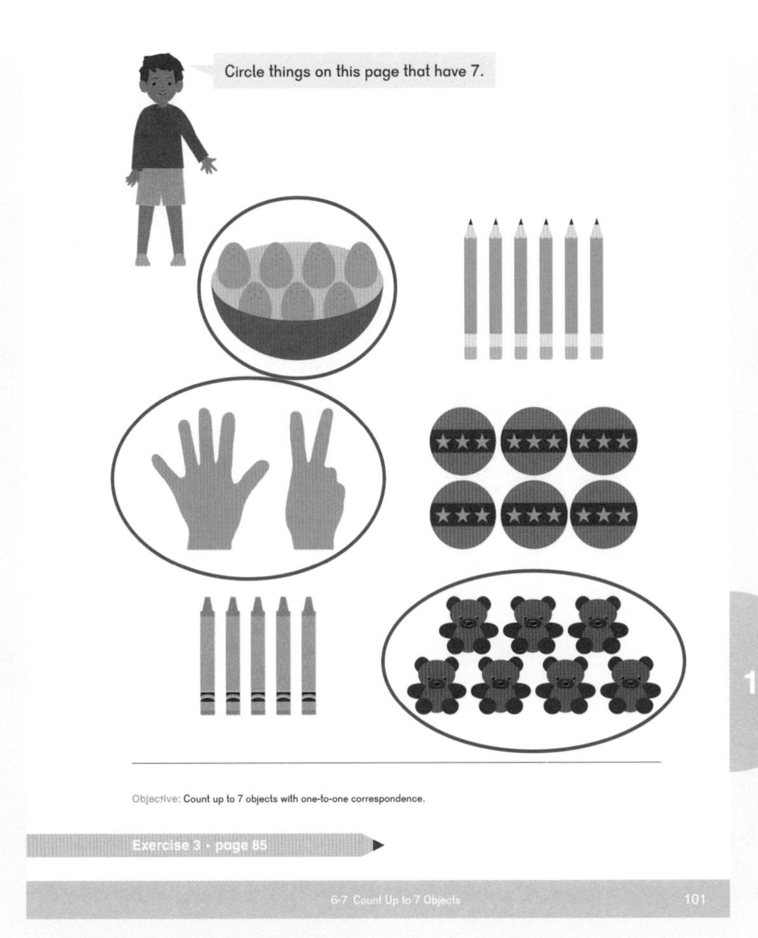

Small Group Center Play

Sort: Include several types of objects, up to seven of each, and have students sort the objects in more than one way.

Office: Include clothes that an adult would wear for dress-up, magazines, and newspapers, and encourage students to copy any numbers they see.

Pretty 7 Butterfly: Have students draw a butterfly and color it with seven colors.

Counting: Provide small objects and 10-divot egg cartons. Have students count out 7 objects and place them in the egg carton as if placing them on a ten-frame, counting each as it is placed. Then have them remove the objects, counting back from 7.

Reading Time: Read *The Very Hungry Caterpillar* by Eric Carle to the students. Give them art paper and crayons to draw what they do during one day.

Whole Group Play

7 in the Hoop: Have students sit in a circle with a hula hoop in the middle. Hand each student a toy. The goal is to put one toy inside the hula hoop until there are seven toys. Have students, one at a time, put a toy in, asking, "How many now?" each time. After the seventh toy is added, ask, "How many toys do we want inside the hula hoop? How many are there now? Do we need any more toys inside the hula hoop?" Remove the toys and repeat.

Materials: Hula hoops, 7 toys

Exercise 3 • page 85

Extend Play

Life Cycles: After reading *The Very Hungry Caterpillar,* talk about the life cycle of a butterfly.

7 Flowers for 7 Butterflies: Have students add to their butterfly art by drawing seven flowers.

Materials: Student butterfly art from **Pretty 7 Butterfly** activity.

Objective

- Count up to 8 objects with one-to-one correspondence.

Lesson Materials

- Paper cups
- Bags containing 10 small counters each, 1 bag per student
- Large ten-frames showing 8
- Optional snack: Octopus snacks using a round cracker and 8 pretzel sticks

Explore

Show a ten-frame card showing eight and have students count the dots with you, emphasizing that the number after seven is eight.

Give each student a bag of counters. Have students count out eight counters and put them into paper cups. Then have them remove the counters, one at a time, counting back from eight as they replace the counters in the bag.

Learn

Have students look at page 102 and identify the creature. Read Dion's comment aloud and count the number of arms with them. Have them identify the number shown on the ten-frame card on the page and the number of spots on the octopus's head.

Whole Group Play

Lead students in exercises, counting each movement, eight of each.

Small Group Center Play

Sort: Provide beans or buttons of different types and have students sort them in more than one way.

Kitchen: Provide baking tins and play cookies. Have students count 8 cookies and pretend to bake them.

Octo-8: Give each student an Octopus Template (BLM) and some stickers. Have each student count out eight stickers and put a sticker on each arm.

Counting: Provide small objects and 10-divot egg cartons. Have students count out 8 objects and place them in the egg carton as if placing them on a ten-frame, counting each as it is placed. Then have them remove the objects, counting back from 8.

Exercise 4 • page 87

Extend Learn

Play 8: Play a full octave and define the word for students as, "Eight musical notes in order." Allow students to play their keyboards and sing.

Materials: Keyboards or Keyboard Templates (BLM)

Objective

- Count up to 9 objects with one-to-one correspondence.

Lesson Materials

- Bags containing linking cubes of 3 different colors, 12 cubes per bag, 1 bag per student
- Optional snack: Cracker Jack®, if possible, otherwise 9 grapes per student

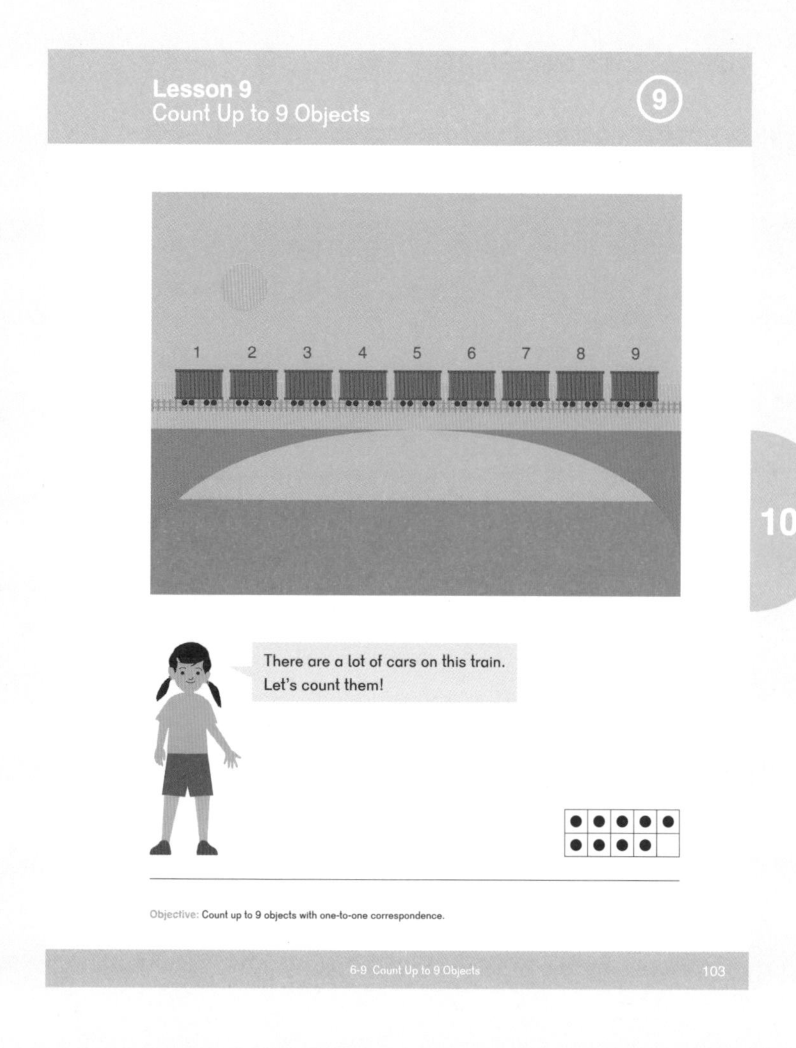

Explore

Give each student a bag of linking cubes and tell them that their task it to make a train using nine linking cubes and three different colors. Challenge them to create a color pattern in their trains.

Have students work in pairs by trading trains and having their partners count the number of linking cubes to make sure nine were used. If trains are made with more or less than nine linking cubes, have students add or remove linking cubes as necessary.

Learn

Have students look at page 103 and read Mei's comment to them. Count the number of train cars with students by putting a finger on each train car as it is counted.

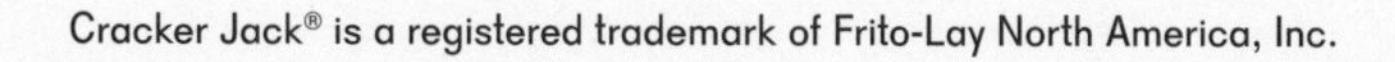

Cracker Jack® is a registered trademark of Frito-Lay North America, Inc.

Have students look at page 104 and discuss the picture. Read Emma's question and ask students to count each object on the page and say how many.

Whole Group Play

T-Ball: Have students discuss what they know about t-ball. If necessary, tell them that there are nine players on a t-ball team. Divide the class into two teams and play a game of t-ball.

Materials: T-ball bat, ball, and bases

Small Group Center Play

Sort: Provide beans or buttons of different types and have students sort them in more than one way.

Neigh or Maa: Have students pretend to be either a horse or a goat.

All Aboard Trains!: Have students discuss the purpose of trains. Provide the construction paper shapes listed in the **Materials**. Have students look at pictures of trains and create a train using glue sticks and the paper shapes.

Counting: Provide small objects and 10-divot egg cartons. Have students count out 9 objects and place them in the egg carton as if placing them on a ten-frame, counting each as it is placed. Then have them remove the objects, counting back from 9.

Exercise 5 • page 89

Extend Explore

Dot Cards 7, 8, 9: Give each student stickers and index cards. Have them create dot cards for six, seven, eight, and nine.

Materials: Index cards, stickers

Objective

- Count up to 10 objects with one-to-one correspondence.

Lesson Materials

- Bags containing 12 small counters, 1 bag per student
- Egg cartons with 10 divots, 1 per student
- Blank Ten-frames (BLM), 1 per student
- Optional snack: Finger sandwiches

Explore

Give each student a bag of counters, an egg carton, and a Blank Ten-Frame (BLM). Have students count out 10 counters from their bags and place them in the egg carton divots, filling in the top row from left to right first, then the bottom row from left to right. Then have them move the counters, one at a time, to their ten-frames, filling in their cards the same way they filled their egg cartons. Collect materials.

Have students count their fingers with you, making sure that they count their left pinkies first. Ask them how many fingers they have. Then call numbers from 1 to 10 and have them show that number of fingers, starting with their left pinkies. If most students are able to do this, call numbers from 1 to 10 and have them show that number of fingers a different way.

Teach students the rhyme from page 106 of the student textbook, having different students act it out.

Learn

Have students look at page 105 and read Alex's comment aloud. Count fingers again.

Have students look at page 106. Read the rhyme and have students identify which friend is doing the different parts of the rhyme.

Whole Group Play

Name That Finger: Have students stand in a circle with their hands behind their backs. Use an unsharpened pencil or a crayon to touch one of each student's fingers, one at a time. Have students identify the finger you touched.[1]

Materials: Unsharpened pencil or crayon

Hen Race: Mark a start line and a finish line. Have students pretend to be hens. When you say, "Go!" students chicken walk to the finish line.

Small Group Center Play

Sort: Provide 10 each of different objects to sort.

Dress-Up: Provide shoes or belts with buckles and have students practice buckling.

Hands Trace: Help students trace around the outside of both of their hands, then decorate them with a ring or a fingernail on each finger.

Counting: Provide small objects and 10-divot egg cartons. Have students count out 10 objects and place them in the egg carton as if placing them on a ten-frame, counting each as it is placed. Then have them remove the objects, counting back from 10.

Exercise 6 • page 91

Extend Play

Name That Finger (with a twist): Tell students to count their fingers, starting with their left pinkies, and that the number said while counting is the number of that finger. Then, touch a student's finger. That student's task will be to say the number of the finger you touched.

[1]Jo Boaler, in her paper "SEEING AS UNDERSTANDING: The Importance of Visual Mathematics for our Brain and Learning" states, "The need for and importance of finger perception could even be a part of the reason that pianists, and other musicians, often display higher mathematical understanding ... The neuroscientists recommend that fingers be regarded as the link between numbers and their symbolic representation."

Objective

- Count up to 10 objects with one-to-one correspondence.

Lesson Materials

- Containers of about 50 small counters, 1 container per 4 students
- Clear plastic cups, 1 per student
- Pretend bowling pins (empty water bottles, small paper cups, etc.)
- Pretend bowling ball depending on the size of the pins (marble, ping pong ball, tennis ball, etc.)
- Optional snack: Round crackers with 3 dots made with cream cheese in the shape of bowling ball holes

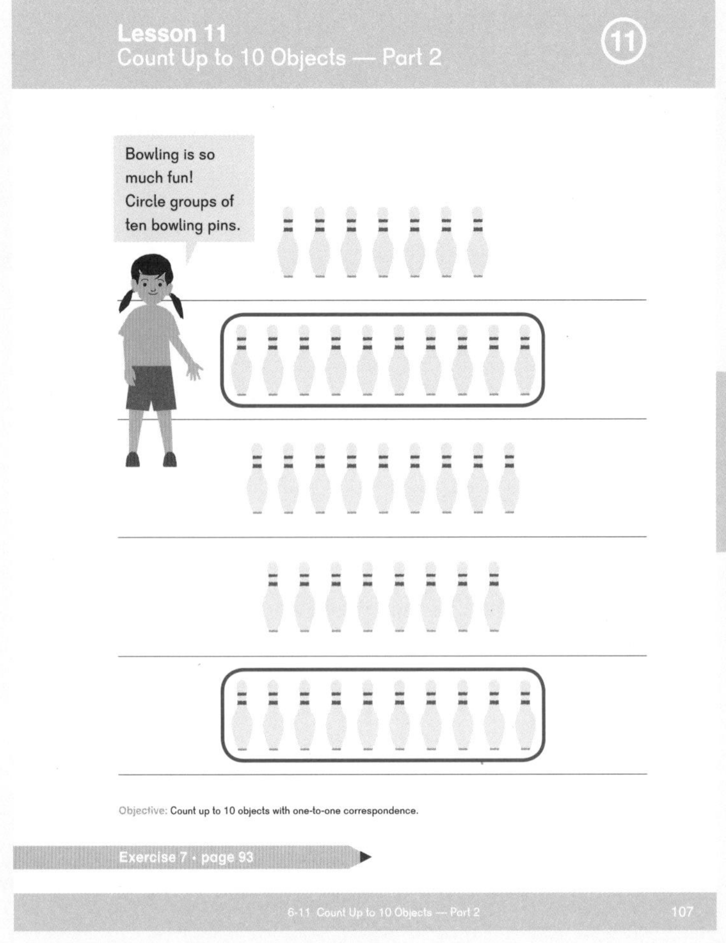

Lesson 11
Count Up to 10 Objects — Part 2
11

Objective: Count up to 10 objects with one-to-one correspondence.

Exercise 7 · page 93

6-11 Count Up to 10 Objects — Part 2 107

107

Explore

Give each student a plastic cup. Put a container of counters near each group of four students. Have each student put 10 counters in their cup, then trade cups with a friend and check counts.

Collect materials and bring student focus back to the whole group. Set up pretend bowling pins as shown below. Have students identify the number of pins.

Explain that the goal in bowling is to roll the ball and knock down as many pins as possible. Demonstrate and have students count with you the number of pins you knocked down.

Learn

Have students look at page 107 and ask them how they would go about counting the pins in each row without double counting. Then show them some methods. For example, cross off pins in the first row with a pencil (or dry erase marker with the page in a sheet protector) as you count them with students. Or, students may place counters on each pin as it is counted.

Read Mei's comment and direction aloud. Have students complete the task row by row, counting pins and circling groups of ten.

Whole Group Play

Catch that Ball: Have up to 10 students stand in a circle and toss a ball to each other.

Materials: Ball

Over Under Relay Race: Have up to 10 students stand in a line. Give a small ball to the first student. That student passes the ball between his or her legs to the second student, who passes the ball over his or her head to the third student, etc. The last student runs to the front of the line with the ball. Repeat until the student who started at the back of the line is back in that position.

Materials: Ball

Small Group Center Play

Sort: Provide 10 each of different objects to sort.

Bowl-a-Rama: Set up bowling pins and have students take turns bowling.

Rock Art: Have students dip rocks into paint and create a 10-frame arrangement of rock prints on a piece of paper.

Counting: Provide small objects and 10-divot egg cartons. Have students count out 10 objects and place them in the egg carton as if placing them on a ten-frame, counting each as it is placed. Then have them remove the objects, counting back from 10.

Exercise 7 • page 93

Extend Play

Catch that Ball and Count: Start with five students playing. If a student misses a toss, another student joins the game. All students then say, "There were five students playing. One more joined. Now there are six of us playing." Repeat until 10 students are playing, then start with a different five students.

Materials: Ball

Objective

- Count up to 10 objects with cardinality.

Lesson Materials

- Bags containing 1 of each of the set of Cuisenaire rods, if possible, otherwise Cuisenaire Rods (BLM), 1 per student
- Crayons
- Optional snack: Pretzel sticks

Explore

Give each student a bag of Cuisenaire rods if possible, or cut-outs from the Cuisenaire Rods (BLM) worksheet. Ask them how many rods they have in their bags.

After counting once, assess cardinality by seeing if they can answer the question without re-counting. Have students order the rods from shortest to longest and ask them again how many rods they have. Review length comparison by having students tell which rod is shorter than another, longest, etc. Collect the materials.

Call out a number and have students show that number of fingers, starting with their left pinkies. Then have them show the same number of fingers a different way.

Learn

Have students look at page 108. Read Alex's question aloud and have them answer. Ask them how many fingers Alex is showing on each hand, then how many fingers he is showing altogether. Repeat for Dion.

Have students look at page 109. Read Dion's question aloud. Have students count the bottom right group of six cans, crossing off each can as it is counted. Have them color in six on the ten-frame card. Read Dion's direction aloud and have them complete the task.

Whole Group Play

Show Me: Call out different numbers 2 to 10 and have students show you that number of fingers any way they choose.

Small Group Center Play

Sort: Provide 10 each of different objects to sort.

Bowl-a-Rama: Set up bowling pins and have students take turns bowling.

Bead Bracelets: Provide beads and pipe cleaners. Have students count out 10 beads and make a bracelet.

Counting: Provide small objects and 10-divot egg cartons. Have students count out 10 objects and place them in the egg carton as if placing them on a ten-frame, counting each as it is placed. Then have them remove the objects, counting back from 10.

109

Exercise 8 • page 95

Extend Play

What's Gone?: Show students sets of up to 10 different small objects, starting with sets of five. First, have them identify how many objects you are showing without counting. Have them close their eyes as you remove an object. Then, have them open their eyes and identify which object is missing. Repeat, adding another object to the original set as students become more skillful.

Materials: Sets of objects in quantities of 5 to 10

Objective

- Practice concepts introduced in this chapter.

Lesson Materials

- Optional snack: Apple slices

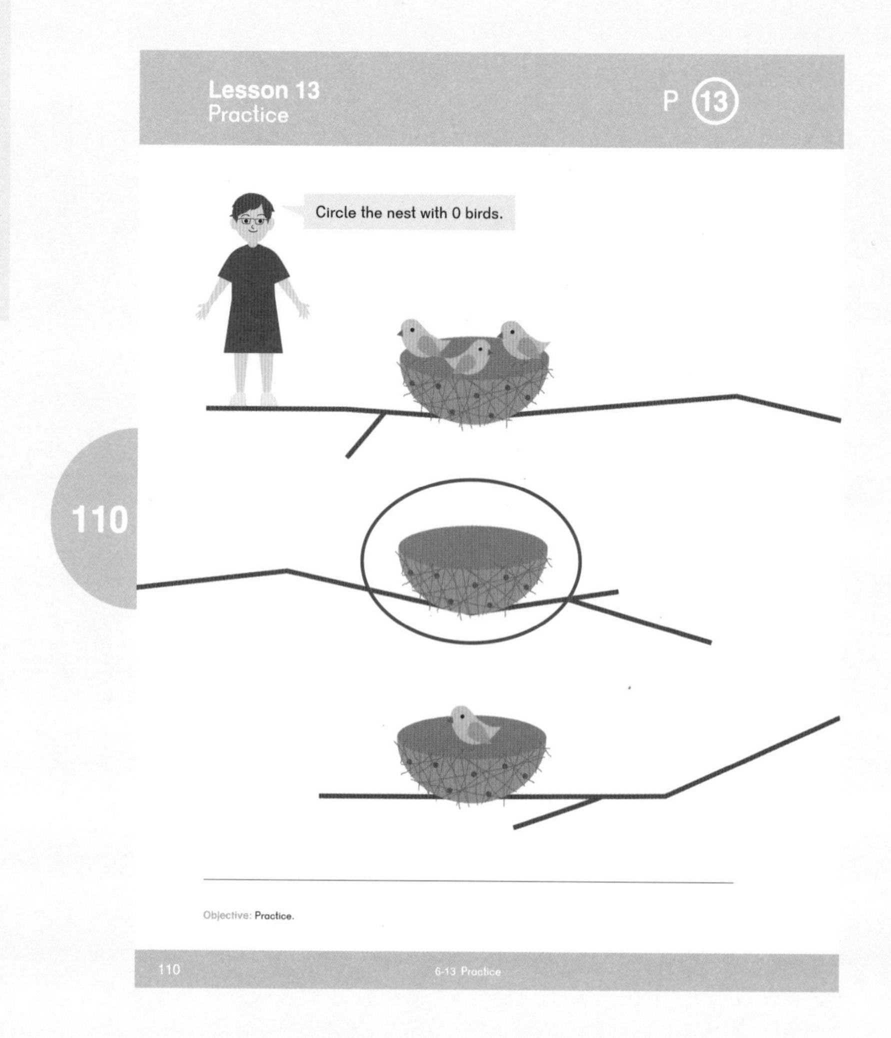

For the **Practice**, read the directions and speech bubbles on each page and have students complete the task.

Small Group Center Play

Prior to sending students to centers, read *Ten Eggs in a Nest* by Marilyn Sadler aloud.

Sort: Plastic eggs into ten-divot egg cartons.

Act it Out: Have students act out the story.

Eggs in a Nest: Include strips of paper in many colors, yellow pom-poms, and googly eyes. Have students glue strips of paper together to create a nest. Have them make chicks out of pom-poms and googly eyes. Then have them choose a number of chicks, including zero, to put in their nests.

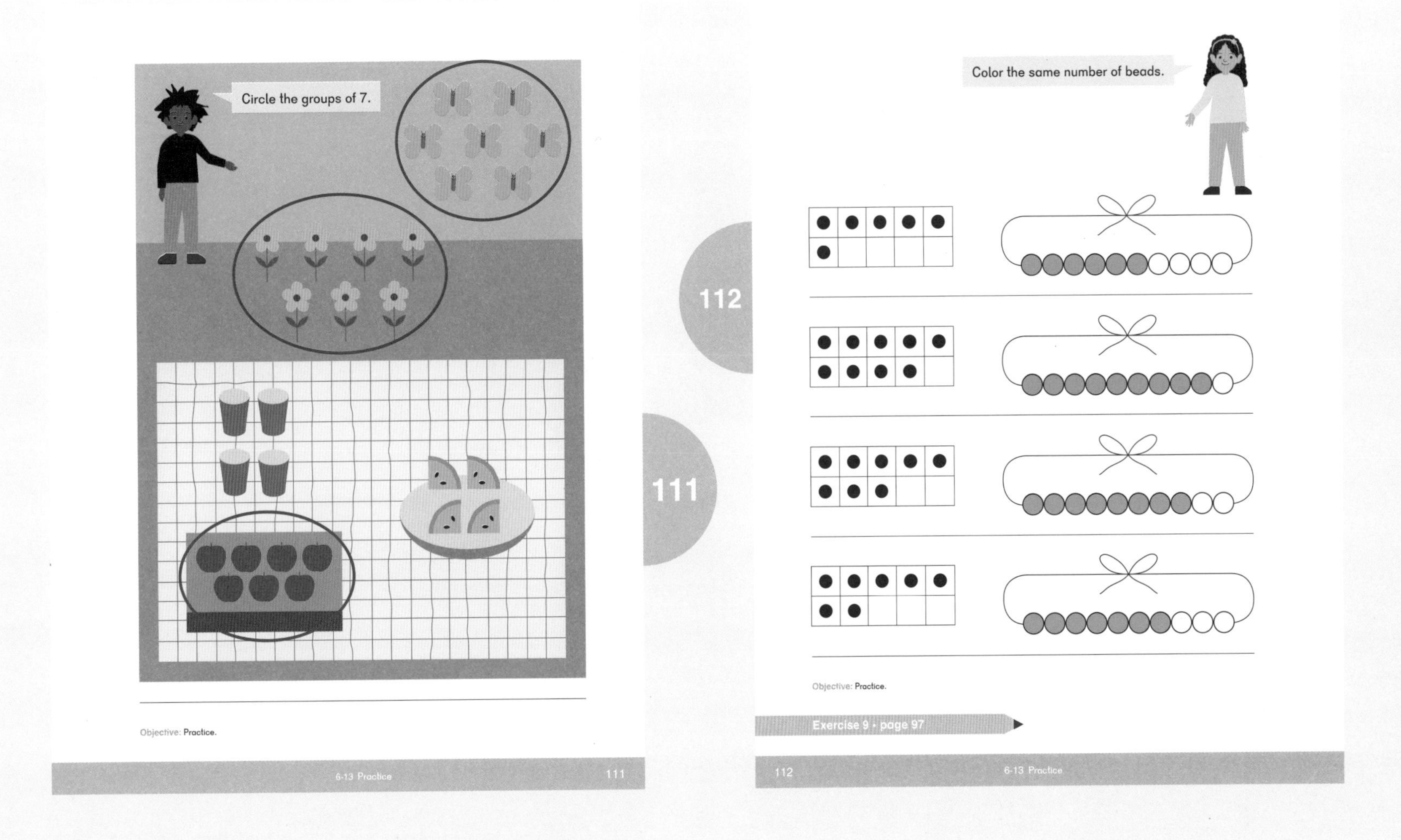

Exercise 9 • page 97

Extend Play

Eggs In a Hole: Have students help make eggs in a hole (sometimes called "Egg in the Basket") using eggs and bread. Enjoy!

Materials: Bread, eggs, toaster oven

Exercise 1 • pages 81 – 82

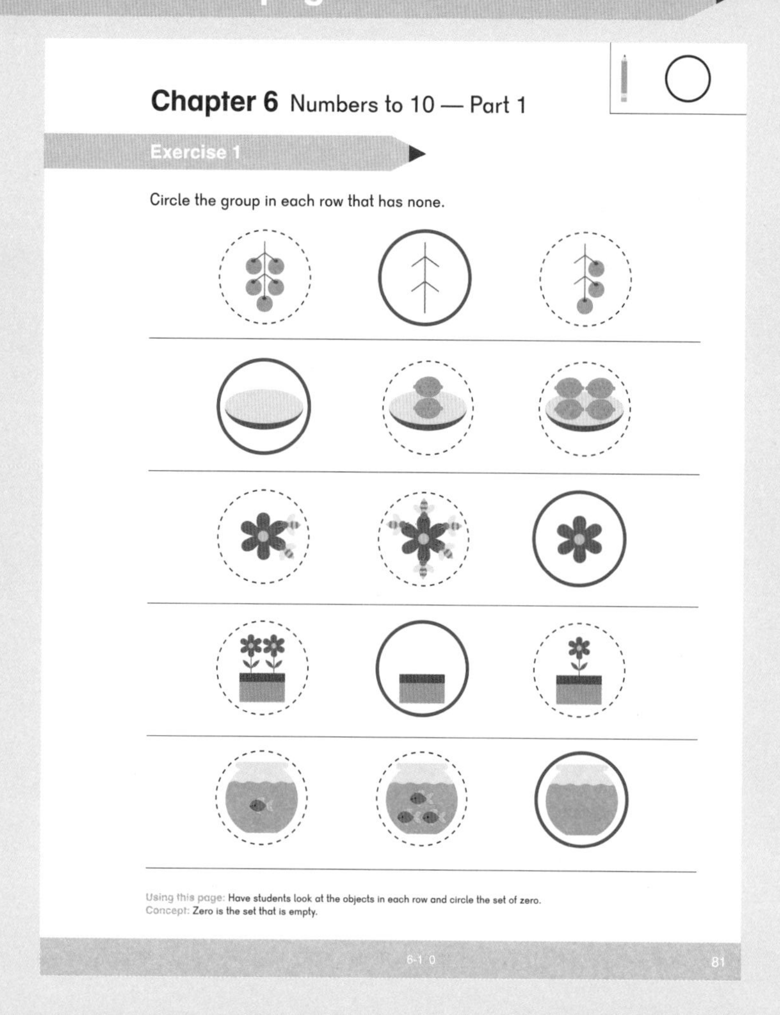

Chapter 6 Numbers to 10 — Part 1

Exercise 1

Circle the group in each row that has none.

Using this page: Have students look at the objects in each row and circle the set of zero.
Concept: Zero is the set that is empty.

6-1 0 81

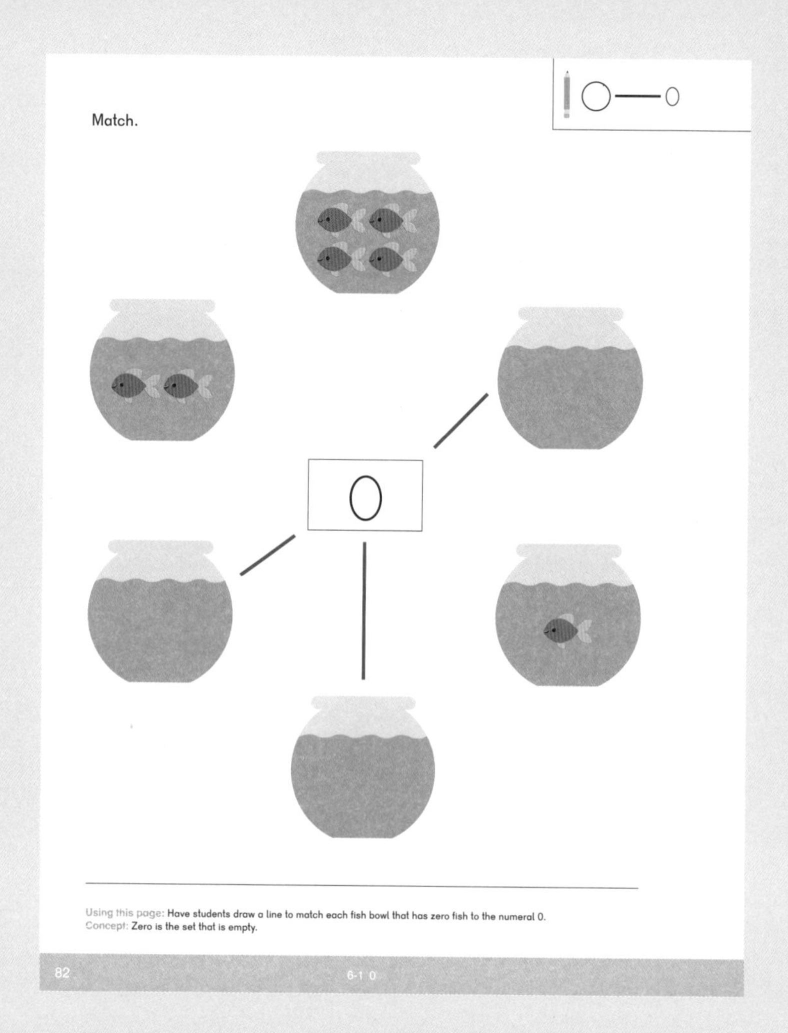

Match.

0

Using this page: Have students draw a line to match each fish bowl that has zero fish to the numeral 0.
Concept: Zero is the set that is empty.

82 6-1 0

Exercise 2 • pages 83 – 84

Exercise 2

Count and circle the groups of 6.

Using this page: Have students count the objects in the picture and circle the 2 sets of 6.
Concept: Counting to 6 with one-to-one correspondence.

6-6 Count Up to 6 Objects 83

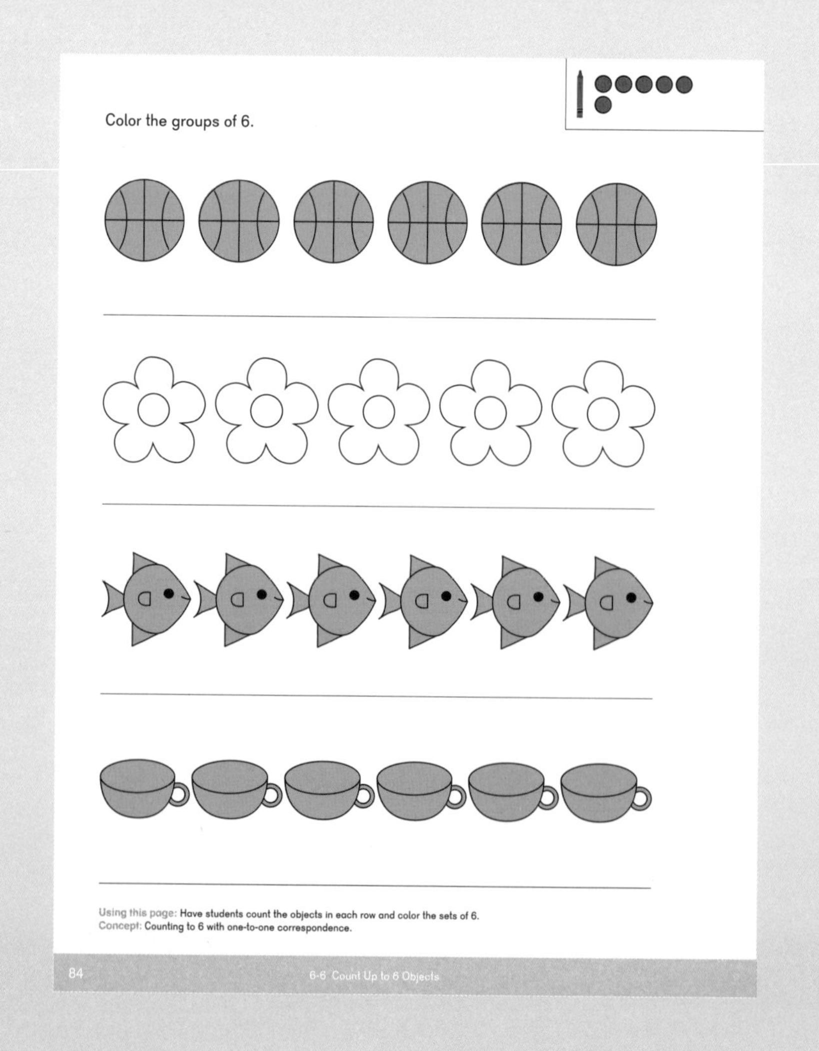

Color the groups of 6.

Using this page: Have students count the objects in each row and color the sets of 6.
Concept: Counting to 6 with one-to-one correspondence.

84 6-6 Count Up to 6 Objects

Exercise 3 • pages 85 – 86

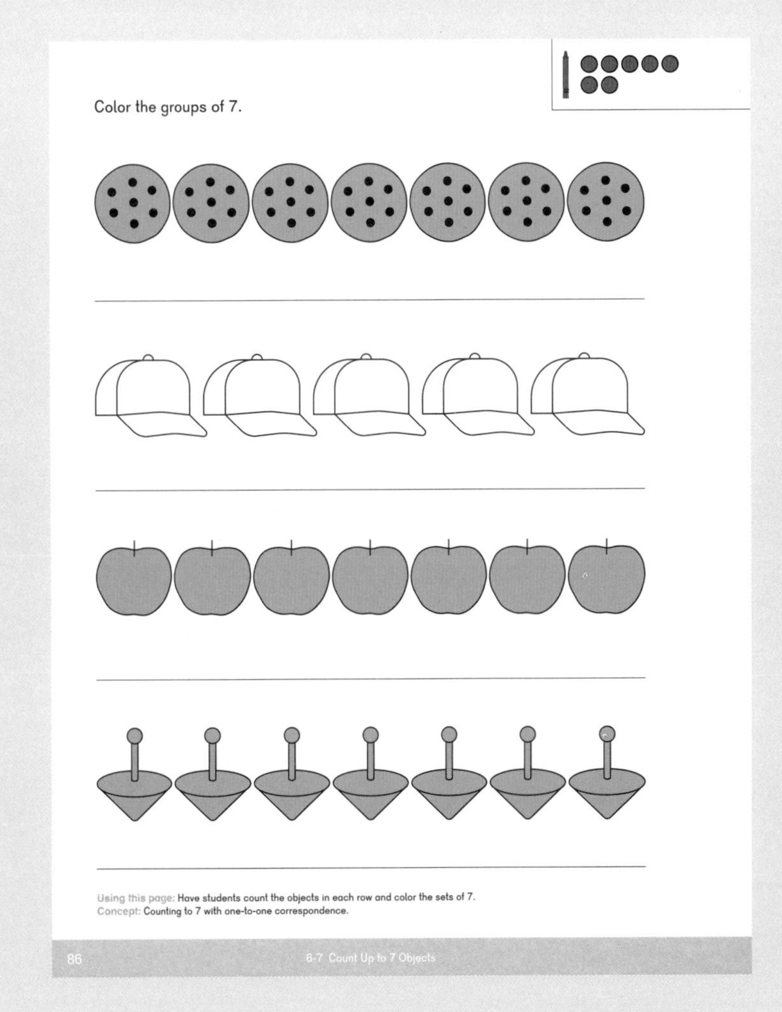

Exercise 4 • pages 87 – 88

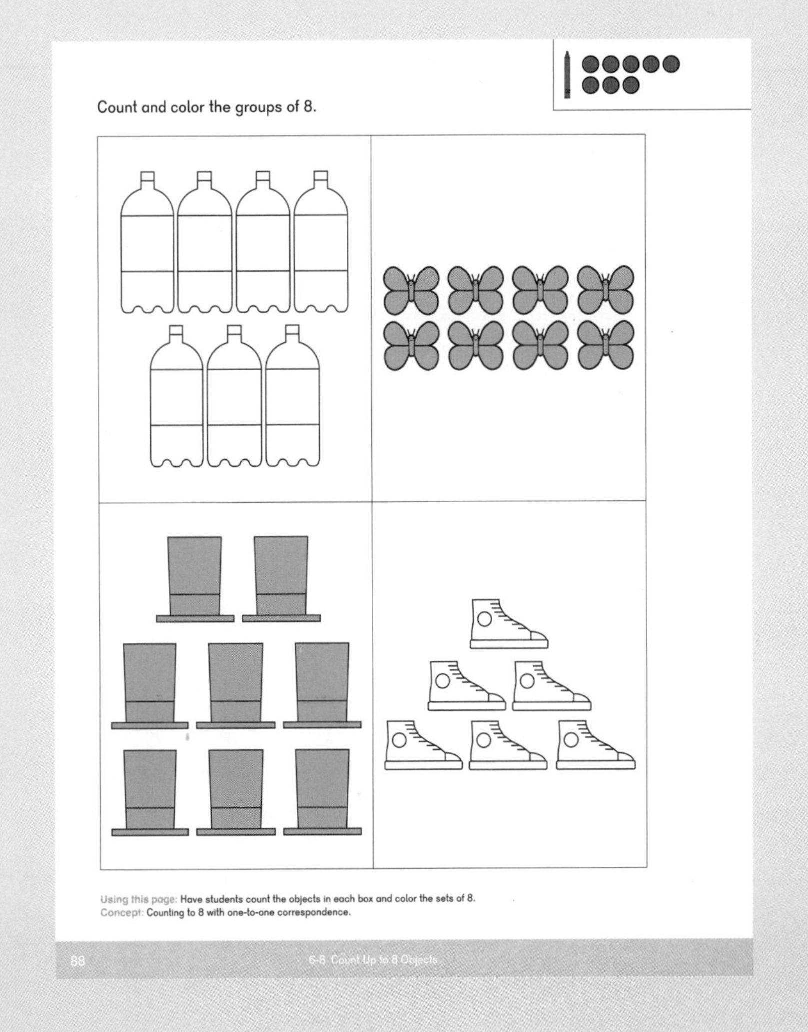

Exercise 5 • pages 89 – 90

Exercise 5

Count and circle the group of 9.

Using this page: Have students count the objects in the picture and circle the set of 9.
Concept: Counting to 9 with one-to-one correspondence.

6-9 Count Up to 9 Objects 89

Count and color the groups of 9.

Using this page: Have students count the objects in each box and color the sets of 9.
Concept: Counting to 9 with one-to-one correspondence.

90 6-9 Count Up to 9 Objects

Exercise 6 • pages 91 – 92

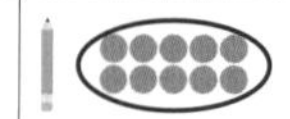

Exercise 6

Count and circle the group of 10.

Using this page: Have students count the objects in the picture and circle the set of 10.
Concept: Counting to 10 with one-to-one correspondence.

6-10 Count Up to 10 Objects — Part 1 91

Paste dot stickers to show 10.

Before using this page: Cut out dots from paper or cut apart sheets of dot stickers and distribute 15 dots to each student.
Using this page: Have students count out 10 dots and paste them onto the wings of the top ladybug. Next, have students count the spots on the bottom ladybug, and count on the remaining stickers till 10, then paste the stickers on the right wing.
Concept: Counting to 10 with one-to-one correspondence.

92 6-10 Count Up to 10 Objects — Part 1

Exercise 7 • pages 93 – 94

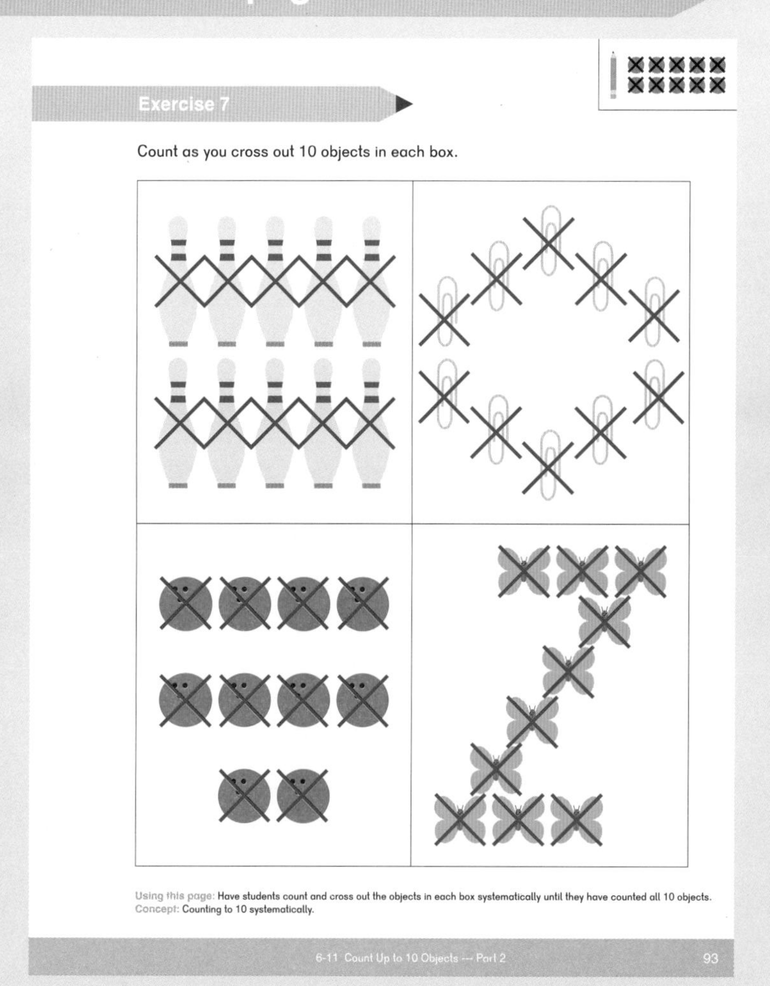
Exercise 7

Count as you cross out 10 objects in each box.

Using this page: Have students count and cross out the objects in each box systematically until they have counted all 10 objects.
Concept: Counting to 10 systematically.

6-11 Count Up to 10 Objects — Part 2 93

Count and color the groups of 10.

Using this page: Have students count the objects systematically in each box, then color the sets of 10.
Concept: Counting to 10 systematically.

94 6-11 Count Up to 10 Objects — Part 2

Exercise 8 • pages 95 – 96

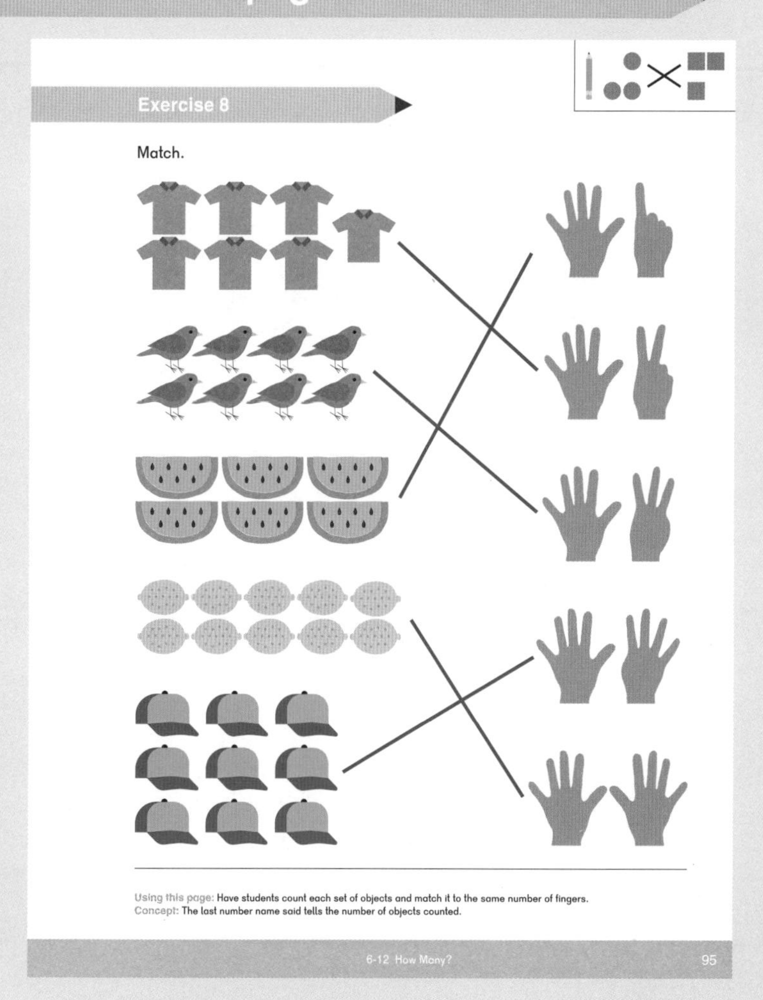
Exercise 8

Match.

Using this page: Have students count each set of objects and match it to the same number of fingers.
Concept: The last number name said tells the number of objects counted.

6-12 How Many? 95

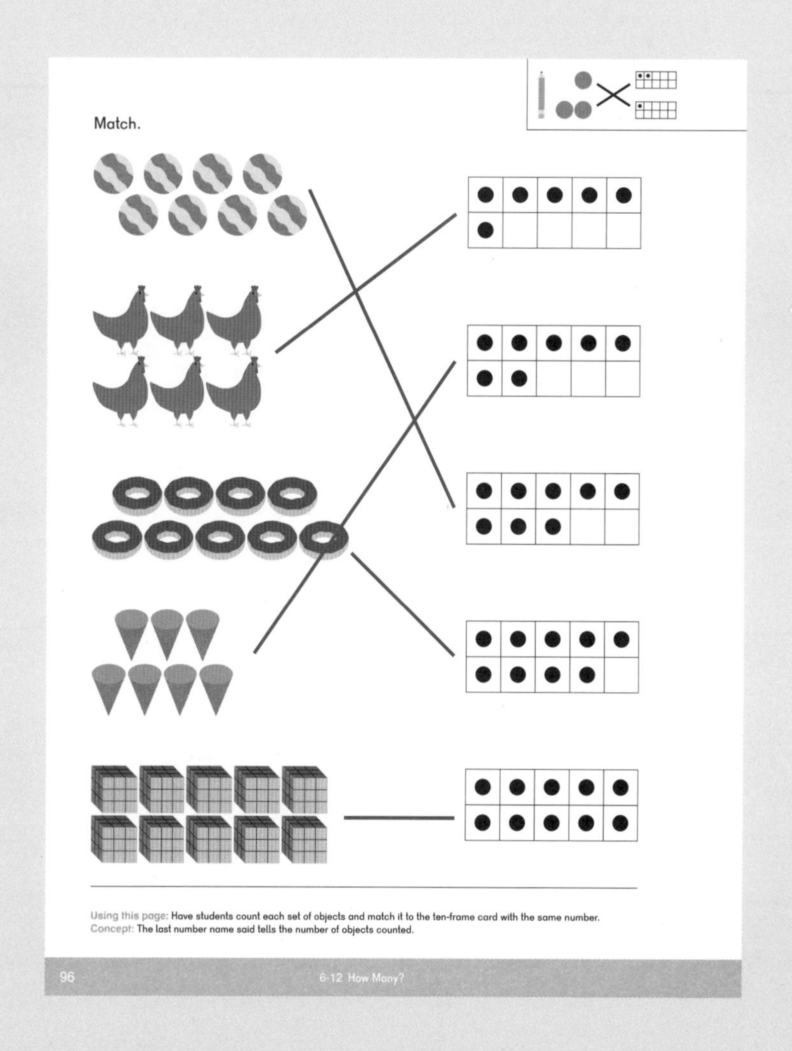
Match.

Using this page: Have students count each set of objects and match it to the ten-frame card with the same number.
Concept: The last number name said tells the number of objects counted.

96 6-12 How Many?

Exercise 9 • pages 97 – 98

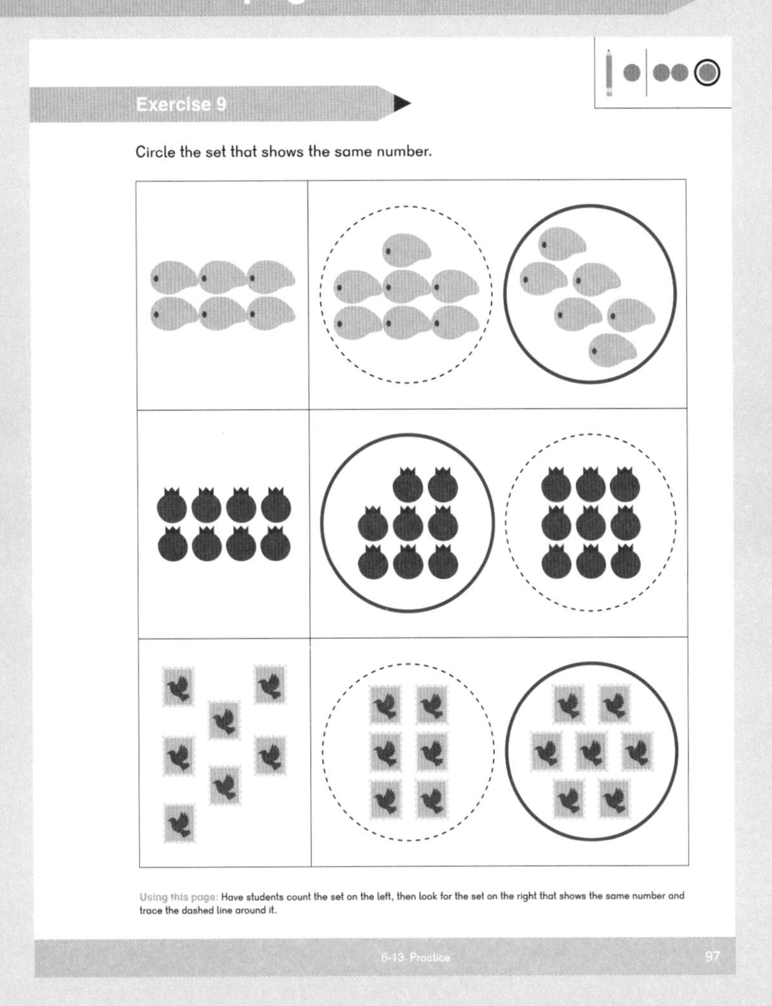

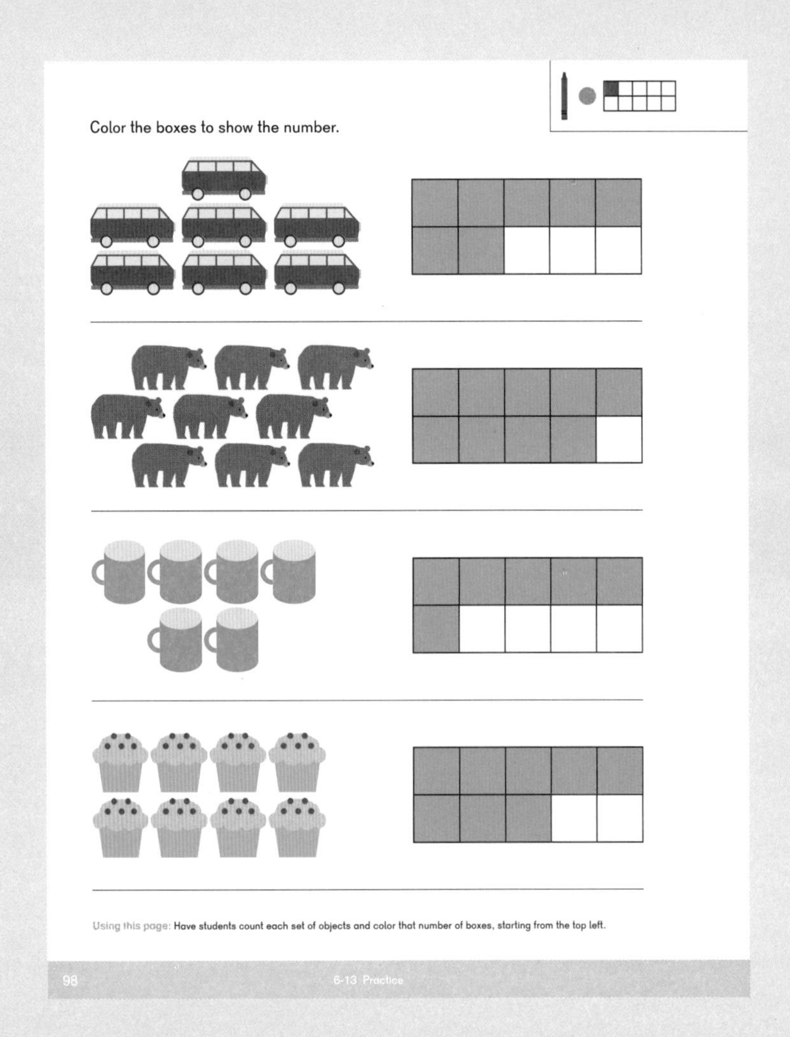

Suggested number of class periods: 9 – 10

Lesson		Page	Resources	Objectives
	Chapter Opener	p. 172	TB: p. 113	
1	6	p. 174	TB: p. 114 WB: p. 99	Recognize the numeral 6.
2	7	p. 176	TB: p. 116 WB: p. 101	Recognize the numeral 7.
3	8	p. 178	TB: p. 118 WB: p. 103	Recognize the numeral 8.
4	9	p. 181	TB: p. 121 WB: p. 105	Recognize the numeral 9.
5	10	p. 183	TB: p. 123 WB: p. 107	Recognize the numeral 10.
6	0 to 10	p. 185	TB: p. 125 WB: p. 109	Review the numerals 0 to 10.
7	Count and Match — Part 1	p. 187	TB: p. 127 WB: p. 111	Match a given numeral 0 to 10 to a set containing that number of objects and to a ten-frame card representing the same.
8	Count and Match — Part 2	p. 190	TB: p. 128 WB: p. 115	Practice matching a given numeral 0 to 10 with a drawn picture containing that number of objects.
9	Practice	p. 192	TB: p. 131 WB: p. 117	Practice concepts introduced in this chapter.
	Workbook Solutions	p. 195		

In **Chapter 4: Numbers to 5 — Part 1** and **Chapter 5: Numbers to 5 – Part 2,** students learned to count by rote, count with one-to-one correspondence, count with cardinality and identify numbers 1 to 5. In **Chapter 6: Numbers to 10 — Part 1,** students learned about 0 and 6 through 10. This chapter builds on prior knowledge by focusing on the numerals 6 through 10 and the ability to match a set of objects containing 0 to 10 to the correct numeral.

Many of the lessons in this chapter call for ten-frame representation of numerals at the bottom of the number cards. This type of card is available in the **Blackline Masters** and identified as "Ten-frame TFR (BLM)" (ten-frame representation). Because the goal of this chapter is to have students recognize all numerals 0 to 10, use this type of number card as much as necessary and differentiate based on the needs of your students. Starting with **Lesson 7: Count and Match,** try to use number cards without the ten-frame representations with all students.

Key Points

Estimation will be introduced in this chapter. This skill is important, but is developmentally challenging for young students. According to John Van de Walle, estimation is a higher-order thinking skill.[1] Continue to practice estimation throughout the year.

Van De Walle J (2006). *Elementary and Middle School Mathematics* (6th ed.). Boston, MA: Pearson.

Materials

- Counters, including teddy bear counters and two-sided counters
- Bags
- Marbles
- Dot stickers
- Linking cubes
- Cuisenaire rods
- Fly swatters
- Packing peanuts
- Pattern blocks
- Egg cartons with 6 divots each
- Egg cartons with 10 divots each
- Beads
- String
- Sand
- Pom-poms
- Cotton balls
- Baskets
- Index cards
- Magazines or newspapers containing pictures of people, animals, or objects
- Balls
- Computers or calculators
- Percussion instruments, kazoos, xylophones and/or keyboards
- Jewelry and superhero attire for dress-up
- Paper towel and toilet paper tubes
- Spoons
- Painter's tape
- Yarn
- White soap (shavings) or powdered soap
- Food coloring
- Wax paper
- Paper plates
- Sand
- Medicine droppers
- Artificial flowers
- Vases (or tall thin containers)

Note: Materials for Activities will be listed in detail in each lesson.

Blackline Masters

- Ten-frame Cards 0 to 10
- Blank Ten-frames
- Number Cards TFR — Small
- Number Cards
- Number Cards — Large
- Dot Cards
- Picture Cards

Storybooks

- *Ten Black Dots* by Donald Crews

Optional Snacks

- "Ants on a Log" (cream cheese or yogurt on celery sticks, topped with raisins)
- Gummy bears
- Grapes
- Pretzel sticks
- Bananas
- Raisins
- Butterfly-shaped crackers
- Rice cakes
- Finger sandwiches
- Animal crackers
- "Flowers" (baby carrots or bell pepper slices and cucumber slices)

Letters Home

- Chapter 7 Letter

Lesson Materials

- "The Ants Go Marching" (VR)
- *Ten Black Dots* by Donald Crews
- Optional snack: "Ants on a Log:" cream cheese or yogurt on celery sticks, topped with raisins

Explore

Read *Ten Black Dots* aloud. Show them only one page at a time for numbers 1, 2, 3, 4, 5, and 7 so that they only see the number of dots that matches the numeral. The representations for 6, 8, 9, and 10 are shown on two pages at a time so show them both pages.

Teach students the song "The Ants Go Marching" (VR) and have them march while singing.

Learn

Have students look at page 113 and have them identify the insects. Then ask them to tell you how many ants are in the first through fifth rows without counting and ask them how they know. Some students may subitize while others may recognize the numerals and assume that represents the number of ants in the row.

Count the ants in the sixth row with students and then ask them how many ants are in the row to assess cardinality. Repeat for the seventh through tenth rows. It is not necessary to teach the ordinal position words "sixth" through "tenth." Those words will be taught in Kindergarten.

Chapter 7

Numbers to 10 — Part 2

1
2
3
4
5
6
7
8
9
10

113

113

Whole Group Play

Marching Ants: Have students line up and march as they sing "The Ants Go Marching."

Materials: Video for "The Ants Go Marching" (VR).

Reading Time: Read *Ten Black Dots* aloud. Show students the pictures in the book showing three dots. Ask them what is alike and what is different about the arrangement of the dots. For the dots with the shoelace, assess early part-whole thinking by listening for students who say something like, "Two and one make three." Repeat with the pictures showing five dots.

Materials: *Ten Black Dots* by Donald Crews

Small Group Center Play

Patterns: Include linking cubes of three different colors. Have students create a pattern of their choice using up to 10 cubes.

Music: Include percussion instruments. Allow students to play as they wish.

Picnic: Set out play food. Have students pretend to be ants at a picnic.

Perfect 10 Art: Provide large index cards and dot stickers. Have students create their own works of art using 10 stickers and crayons. After finishing their creations, have them count their dots with a partner.

Extend Learn

Growing Pattern Recognition: Ask students if they notice a pattern in the number of ants in each row on page 113. Some may notice that there is one more ant in each row than in the previous row. See if any students know the number 11 by asking them how many ants would be in the row that comes after the last one.

Lesson 1 6

Objective

- Recognize the numeral 6.

Lesson Materials

- Egg cartons with 6 divots each – 1 per student
- Bags containing 10 teddy bears or other counters – 1 bag per student
- Number Cards — Large (BLM) 0 to 6
- Number Cards TFR — Small (BLM) 0 to 6 – 1 set per student
- Dot Cards (BLM) 1 to 6
- Optional snack: Gummy bears or grapes

Explore

Give each student an egg carton and a bag of counters. Have them count six counters and put one in each divot of the egg carton. Then have them count back as they place the counters in the bag.

Show students Number Cards — Large (BLM) for 0 through 5, not in numerical sequence, and have them identify the number. After this review, show them a Number Card — Large (BLM) 6. Ask them to describe the numeral to a partner. Some students may notice that there is a circle at the bottom.

Give each student a set of Number Cards TFR — Small (BLM) 1 to 6. Have them put their cards in order from least to greatest. Collect the materials.

Some students may need to rely on their rote counting rather than cardinality to sequence the numbers. If so, encourage them to count aloud to find the next card in the sequence.

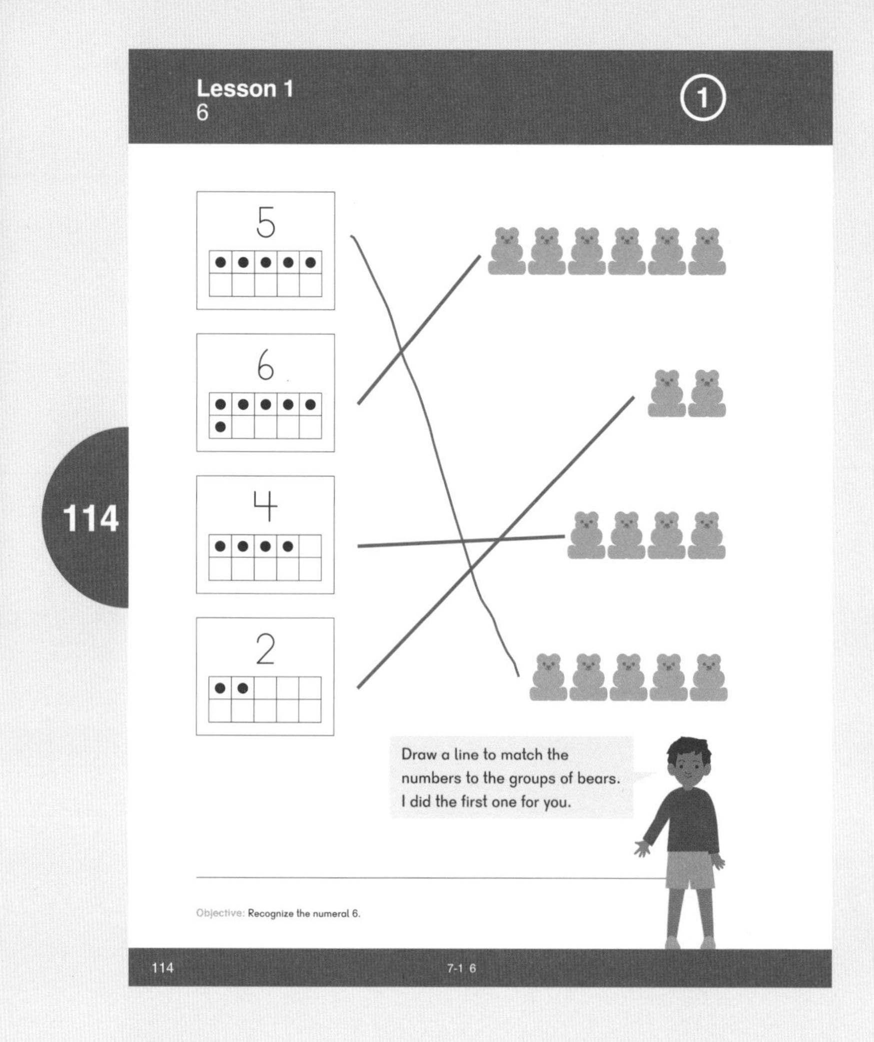

Learn

Have students look at page 114 and identify the objects on the page. Read Alex's direction aloud and have them complete the task.

Have students look at page 115. Read Dion's direction aloud and have them complete the task.

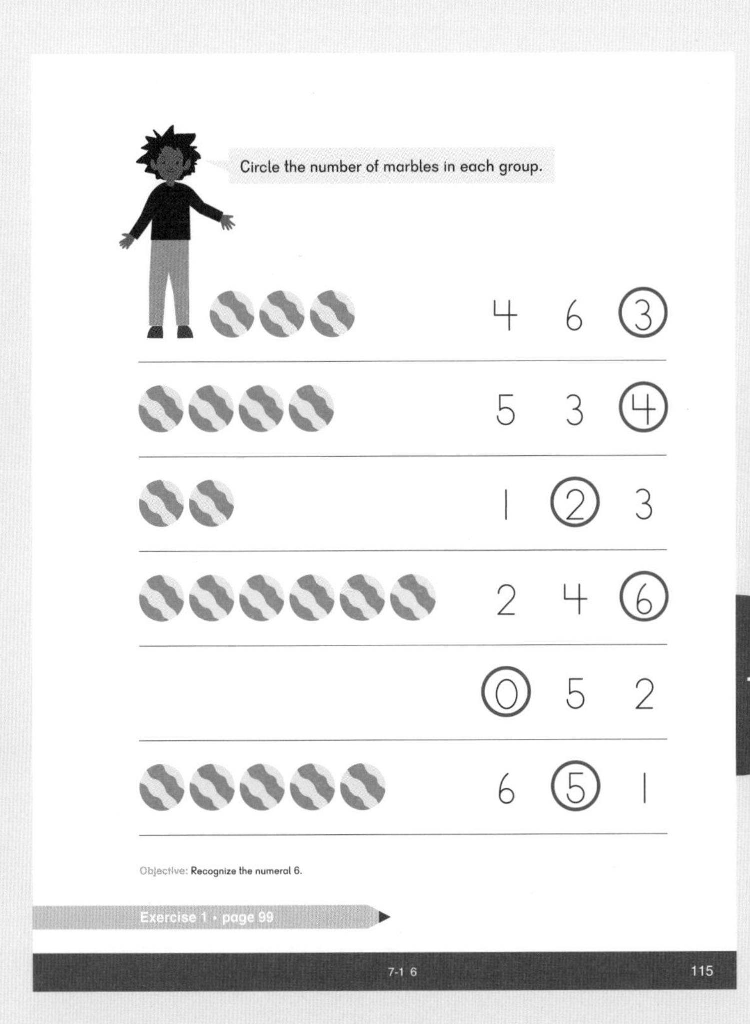

115

Small Group Center Play

Patterns: Include linking cubes of two different colors. Have students create an AB pattern of their choice using six cubes.

Music: Include percussion instruments. Have students create AB sound patterns with them. If possible, record some of their sound patterns and use the recordings to provide a beat for doing movements another time.

Bakery: Set out small bags and play cookies. Have students guess how many cookies will fit in a bag (up to six), then count cookies as they place them in a bag to prove or disprove their theories. Have them show the number card (BLM) for the number of cookies they counted.

Marble Art: Give each student a numeral 6 template (cut out earlier). Have them put six marbles that have been in six paint colors on the template in a tray and roll them around by tipping the tray back and forth.

Whole Group Play

Up to 6 Hopscotch: Use chalk to create a hopscotch court from 1 to 6 and have students sit around it. Give each student a set of number cards like those used in **Learn**. Choose one student to go first. Have that student stand outside the hopscotch court. Call a number from 0 to 6. Standing student must hop on that number while seated students hold up the correct number card. Repeat with a different student until all students have had a chance to hop.

Materials: Chalk or painter's tape, marker for each student

Marble Race: Use paper towel tubes as chutes. Mark a start and a finish line. Have students hold their "chutes" at the start line, drop marbles through it, and see whose marble gets closest to the finish line.

Materials: Paper towel tubes, marbles

Exercise 1 • page 99

Extend Learn

Dot Card Flash: Flash Dot Cards (BLM) showing up to six and ask how many dots students see without counting. Ask them how they knew there were ______ dots each time.

Materials: Dot Cards (BLM) 0 to 6

Objective

- Recognize the numeral 7.

Lesson Materials

- Bags containing 10 beans or other small counters, 1 bag per student
- Number Cards — Large (BLM) 0 to 7
- Number Cards TFR — Small (BLM) 1 to 7 – 1 set per student
- Optional snack: Pretzel sticks joined to make 7s

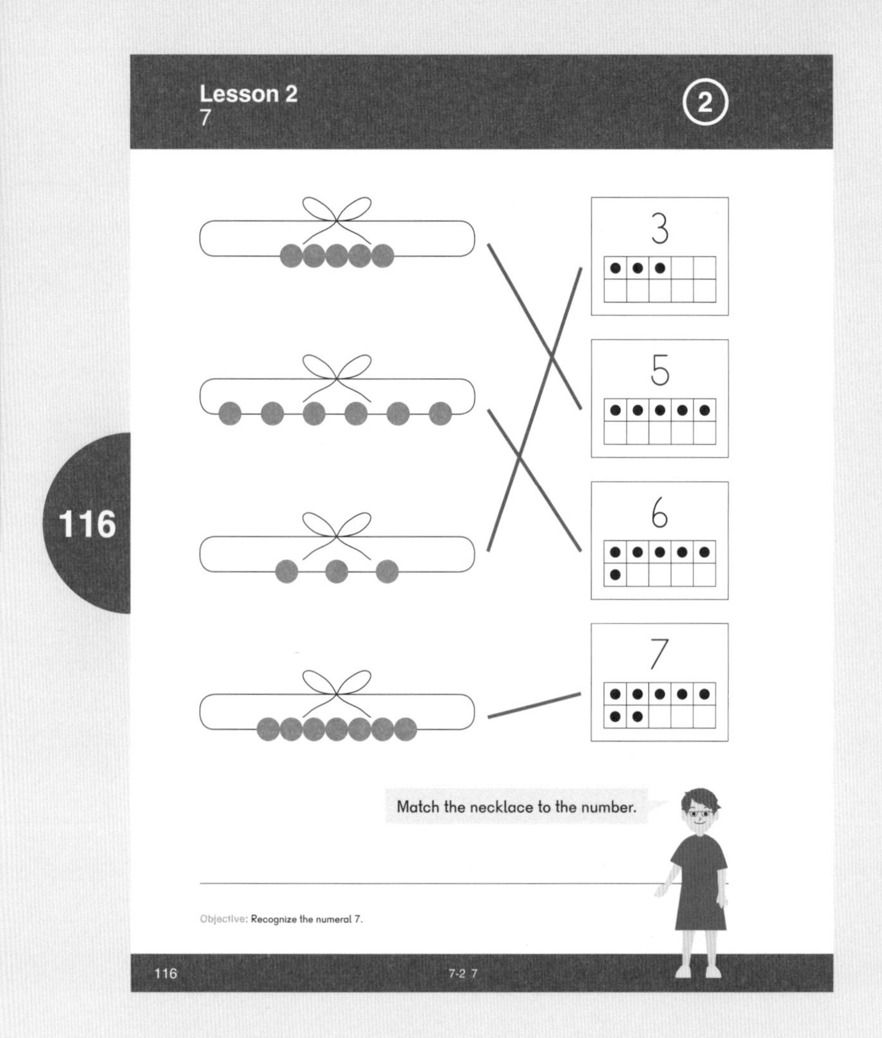

Explore

Show students Number Cards — Large (BLM) for 0 through 6, not in numerical sequence, and have them identify each number. After this review, show them a Number Card — Large (BLM) 7. Ask them to describe the numeral to a partner. Some students may notice that there are two straight lines and no curved parts. Ask which other numbers they have learned about are made of straight lines (1, 4), are curvy (0, 3, 6), and are both (2, 5).

Give each student a set of Number Cards TFR — Small (BLM) and a bag of counters. Have them hold up their 7 cards, count out seven counters and arrange them on the card correctly. Then have them put their cards in order from least to greatest. Collect the materials.

Learn

Have students look at page 116 and read Emma's direction to them. Have them complete the task. Have students look at page 117. Ask them to point to numbers they see on the page and tell a friend which numbers they see and how many of each. Read Sofia's direction and have them complete the task.

Whole Group Play

Find the Number: Have students close their eyes while you post the sticky notes around the room. When students open their eyes, call a number 0 to 7. Students' task will be to find one, and only one, sticky note showing that numeral. Allow students to help each other. Send about five students at a time.

Materials: Sticky notes labeled with numbers 0 to 7

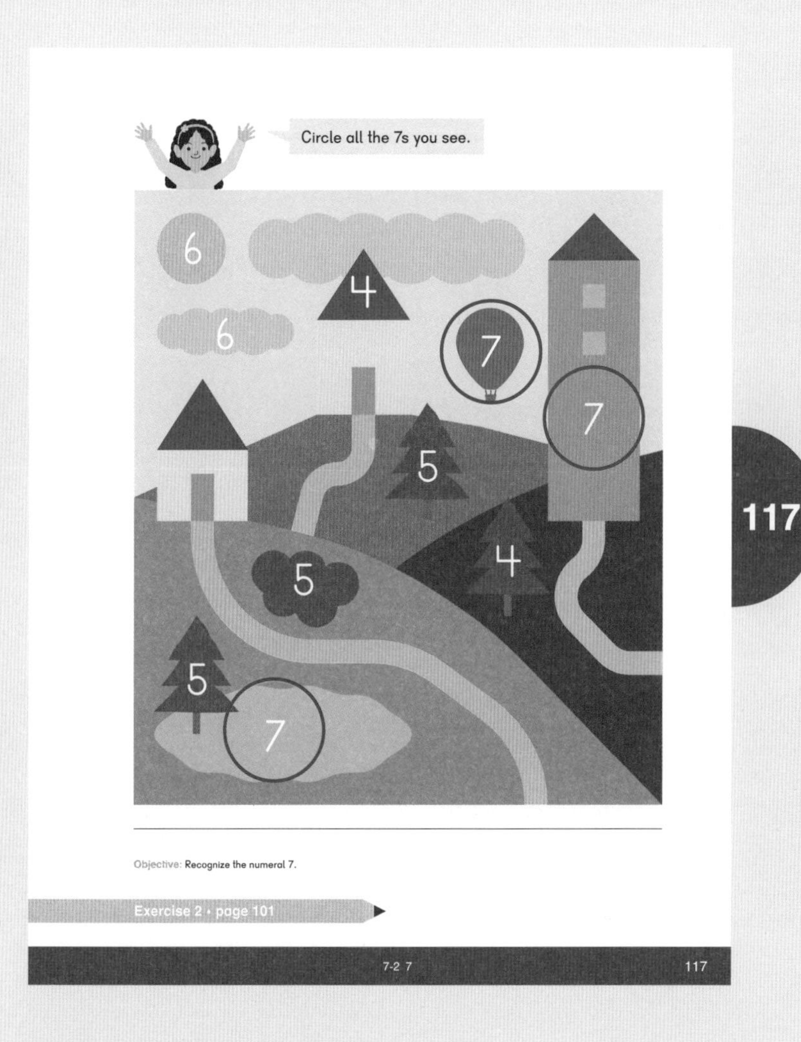

117

Exercise 2 • page 101

Extend Learn

Make 7: Have students color Blank Ten-frame (BLM) templates using two different colors to show different ways to make seven.

Materials: Blank Ten-frames (BLM), crayons

Small Group Center Play

Patterns: Include Cuisenaire rods. Have students create patterns of their choice using up to 10 rods. Then have them use two rods to make the number 7 by laying one rod horizontally and one rod diagonally down from the horizontal rod.

Music: Include spoons, kazoos, and percussion instruments. Have students create their own sounds.

Dress-Up: Include different types of jewelry. Have students use a total of seven things for dress-up.

Bead Art: Have students count seven beads and put them on pipe cleaners. Make bracelets for students to take home.

Sand 7s: Include pieces of art paper with a numeral 7 written on them, sand, and glue sticks. Have them use glue and sand to make sand art on the numeral.

Objective

- Recognize the numeral 8.

Lesson Materials

- Blank Ten-frames (BLM) – 1 per student
- Crayons – 1 per student
- Number Cards — Large (BLM) 7 and 8, Number Cards — Large (BLM) 0 to 8
- Number Cards TFR — Small (BLM) 1 to 8, 1 set per student
- Optional snack: Spiders made with banana slices (body), raisins (eyes), and pretzel sticks (legs)

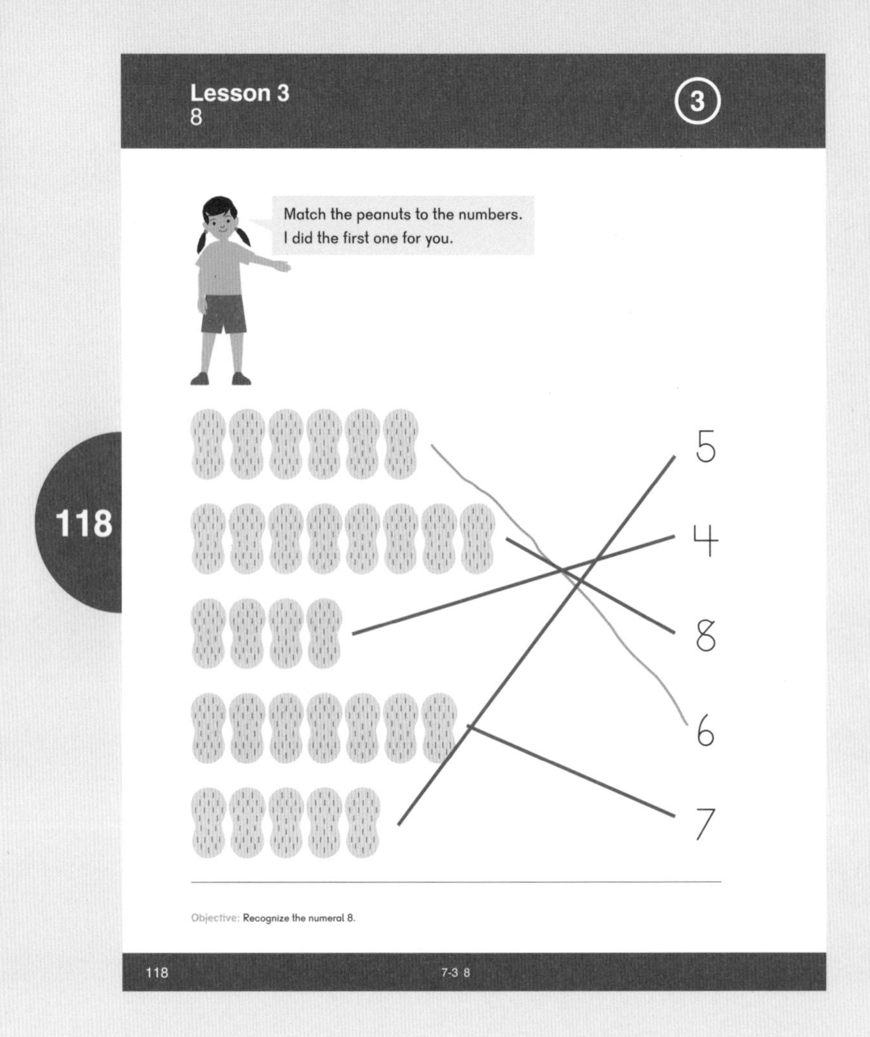

Explore

Give each student a Blank Ten-frame (BLM) and a crayon. Have them use their crayons to show seven on their ten-frames. When they are done, have each look at a friend's ten-frame and make sure he or she showed seven. Then have them trade crayons and use a different color to color one more on their ten-frames. Have them discuss what number they are showing now.

Show students a Number Card — Large (BLM) 8. Ask them to describe the numeral to a partner. Some students may notice that it looks like two circles. Have them compare the numeral 8 to other numerals they have learned about.

Give each student a set of Number Cards TFR — Small 1 to 8 and have them hold up their number cards for 8. Discuss the appearance of the ten-frame for eight. There are five dots colored on the top row and three on the bottom row.

Learn

Have students look at page 118. Tell them that squirrels like to eat peanuts. Read Mei's direction aloud and have them complete the task.

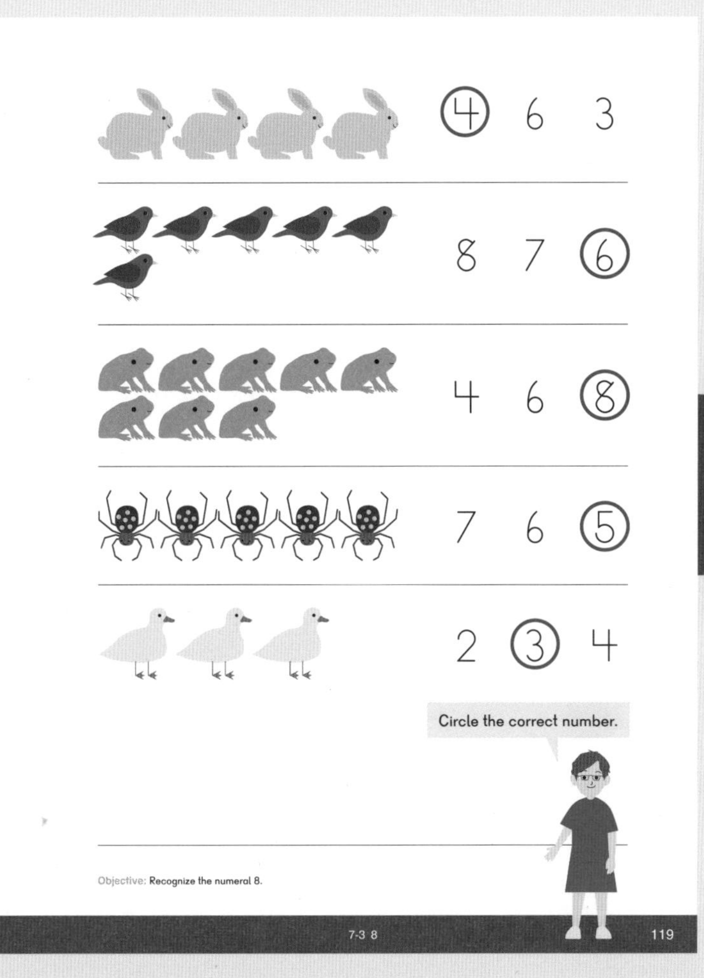

Have students look at page 119 and identify the animals. Ask them how many bunnies there are. See how many students answer correctly without counting. Repeat for the ducks and the spiders. Then ask about the birds and the frogs. If students answered correctly, ask them how they knew. Have them count and say the number of legs on a spider. Read Emma's direction aloud and have them complete the task.

Have students look at page 120 (shown on following page). Read Emma's direction aloud. Point out that a 4 is written under the first column and four boxes are colored. Have them complete the task.

Whole Group Play

That's My Number!: Have students sit on chairs arranged in a circle. Give each student a Number Card (BLM) 0 to 8. Tell students that as you call a number, any students holding that number card must stand up, hold their card up in the air, shout, "That's my number!" and sit down. After all numbers 0 to 8 have been called once, have students exchange cards and repeat. This time, repeat some numbers for example, 1, 3, 2, 2, 2, 4, 5, 8, 8, 6, 1, 7, 7, etc.

Materials: Number Cards (BLM) 0 to 8

Small Group Center Play

Pattern Strips: Include long and short paper strips of the same color. Have students use eight strips to make a pattern of their choice. Be sure to take this opportunity to review long and short from Chapter 1.

Music: Include xylophones and/or keyboards. Have students play a scale and say how many notes they played.

Skate 8: Make a large 8 on the floor with painter's tape. Have students pretend to be ice skaters and skate over the number.

Peanut Creatures: Include packing peanuts and rounded toothpicks. Have students make creatures using eight peanuts.

Exercise 3 • page 103

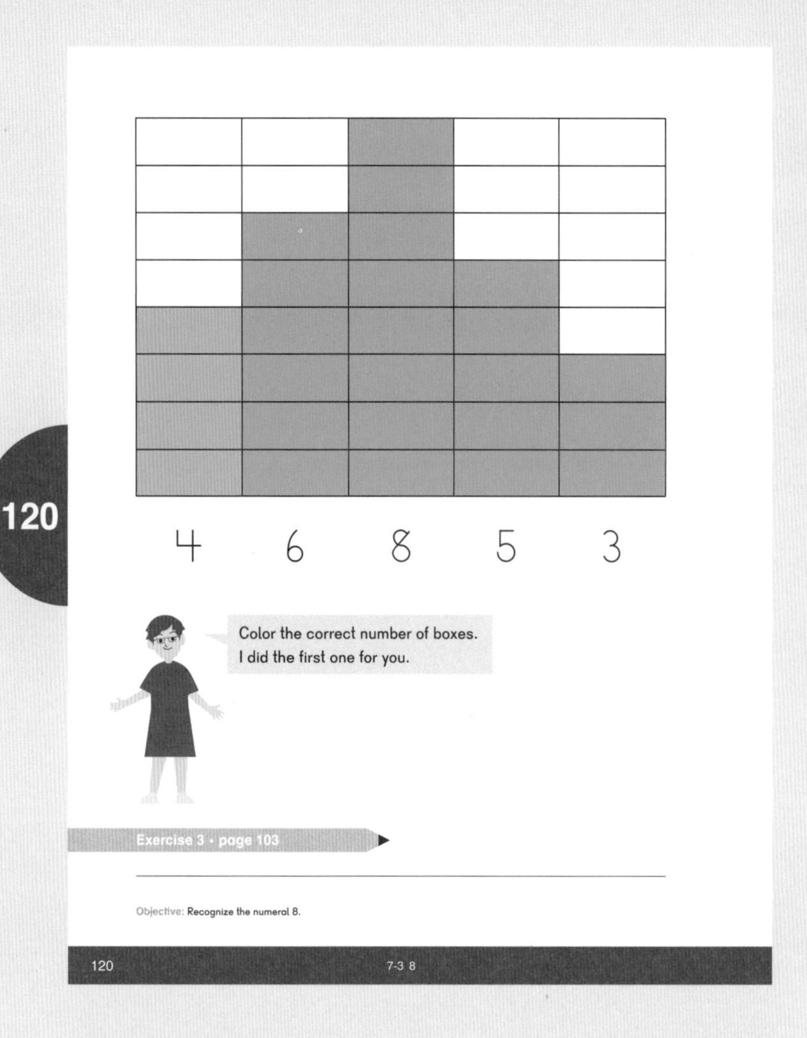

Extend Play

That's My Number! (with a twist): Play the game described in **Play**, but instead of calling a number, give a hint for the number, such as, "The number of eyes I have," or, "The number I say right after five when I'm counting on."

Extend Learn

How Many Shoes?: Pose this problem to the students: "In my closet I saw four pairs of shoes. How many shoes were in my closet altogether?" Allow students to use counters to help them answer.

Materials: Counters

Objective

- Recognize the numeral 9.

Lesson Materials

- Bags containing 10 small counters – 1 bag per student
- Number Cards — Large (BLM) 0 to 9
- Number Cards TFR — Small (BLM) 1 to 9 – 1 set per student
- Blank Ten-frames (BLM)
- Optional snack: Butterfly-shaped crackers

Explore

Show students Number Cards — Large (BLM) for 0 through 8, not in numerical sequence, and have them identify each number. After this review, show them a number card 9. Ask them to describe the numeral to a partner. Some students may notice that there is a loop and a straight line. Show a 6 next to the 9 and have students describe similarities and differences.

Give each student a set of Number Cards TFR — Small (BLM) 1 to 9, a Blank Ten-frame (BLM), and a bag of counters. Have them hold up their 9 card, count out nine counters, and place them on their ten-frames. Then have them put their cards in order from least to greatest. Collect the materials.

Note: The number 9 in the student textbook has been chosen specifically for students. It is a different font than that used in the Teacher's Guides. Have students look for other 9s in the classroom and compare the appearance of each one.

Learn

Have students look at page 121. Read Emma's comment aloud. Have them point to all the 9s on the page, say what number the ten-frame card shows, and how many 9s are on the page.

Whole Group Play

Move 9 Times: Play music and have students do nine toe touches, nine shoulder touches, etc., to the beat of the music.

Materials: Music with a clear beat

Small Group Center Play

Patterns: Include pattern blocks and allow students to create patterns of their choice using nine pattern blocks.

Music: Allow free exploration.

Dressing to the Nines: Provide paper on which 9s have been written, stickers, crayons, scissors, and glue. Have them decorate a 9 as if they were going to glue it to a dress, like Emma did. Then have them glue the 9 to textbook page 121.

Textbook: Have students look at page 122. Read Alex's directions aloud and have them color the butterfly correctly.

122

5 6 7 8 9

Color the 5s blue, the 6s orange, the 7s yellow, the 8s green, and the 9s red.

Objective: Recognize the numeral 9.

Exercise 4 • page 105

122 7-4 9

Exercise 4 • page 105

Extend Play

Nine Rhyme: Before this lesson, prepare art paper by drawing a large 9 on each piece. Have students think of as many words as they can that rhyme with "nine." (Examples: dine, fine, line, sign, spine, shine.) Then give them crayons and a piece of the prepared art paper and have them draw a picture, using the numeral 9, that shows a picture of one of the rhyming words.

Materials: Art paper, crayons

Objective

- Recognize the numeral 10.

Lesson Materials

- 20 baskets
- 110 pom-poms – all the same color, or 110 large cotton balls
- Egg cartons with 10 divots – 1 per pair of students
- Bags containing more than 10 counters – 1 per student
- Number Cards (BLM) 1 to 10, 1 set per pair of students
- Optional snack: Rice cakes decorated to look like kitten faces

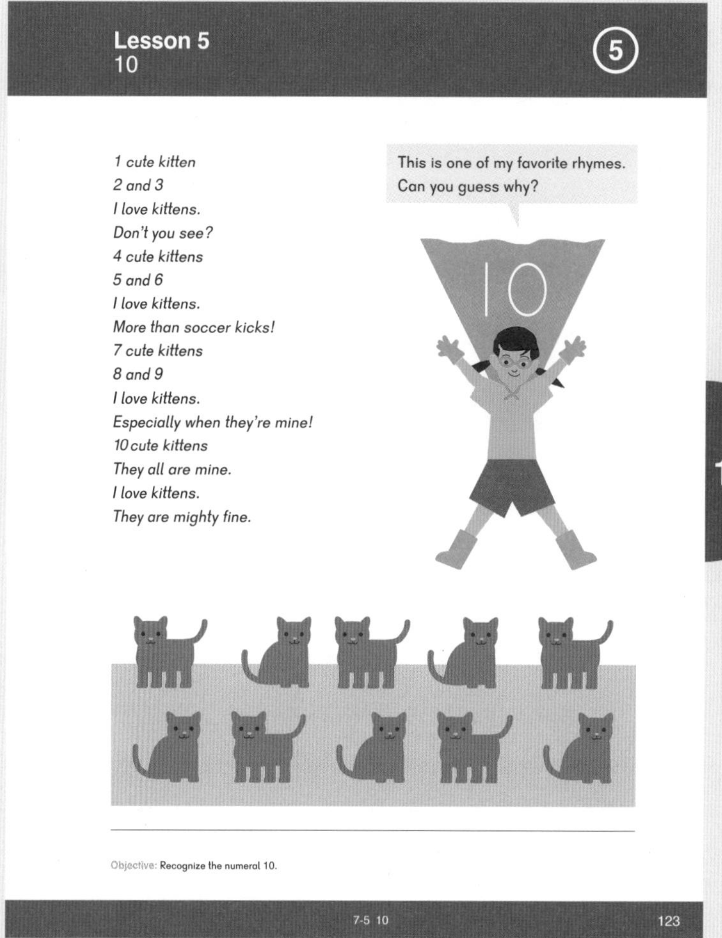

Prior to beginning this lesson, set up two sets of baskets with pom-poms. Have one basket hold one pom-pom, another hold two, etc., up to 10.

Explore

Give each pair of students an egg carton. Give each student a bag of counters. Have students in each pair take turns counting out 10 counters and placing one counter in each divot. Collect the materials.

Show students a Number Card — Large (BLM) for 10 and have them describe the numeral to their partners. Give each pair of students a set of 3 Number Cards (BLM) 1 to 10. Tell them that today they are going to pretend that a pom-pom is a sleeping kitten. When you give the signal, they are going to visit each basket, count the number of sleeping kittens there, and leave the matching number card.

Send 6 or 8 students at a time, utilizing both sets of baskets. As each pair returns, send another pair. While students are waiting, have them put their number cards in order. After all students have visited the baskets, lead a gallery walk around the two sets of baskets. If any baskets have different number cards, count the kittens and discuss which cards should be there.

Learn

Have students look at page 123 and say how many kittens there are. Read the rhyme to the students. Read Mei's comment and have students answer the question. Read the rhyme again and have students count the kittens on the page by touching each time a number word is said. Do this several times and then ask after each time, "How many kittens?"

Have students look at page 124 and say how many buttons are in the first row. Read Dion's direction and have students circle the 5. Then have them complete the task.

Whole Group Play

Swat the Number: With students sitting in a circle, place Number Cards — Large (BLM) 0 to 10 on the floor, two of each. Give each student an individual number path to swat with a finger while they are sitting. Choose two students and give each a fly swatter. Call out a number 0 to 10. All students must swat the number. Continue until all students have had a chance with a fly swatter.

Materials: Fly swatters, Number Cards — Large (BLM) 0 to 10

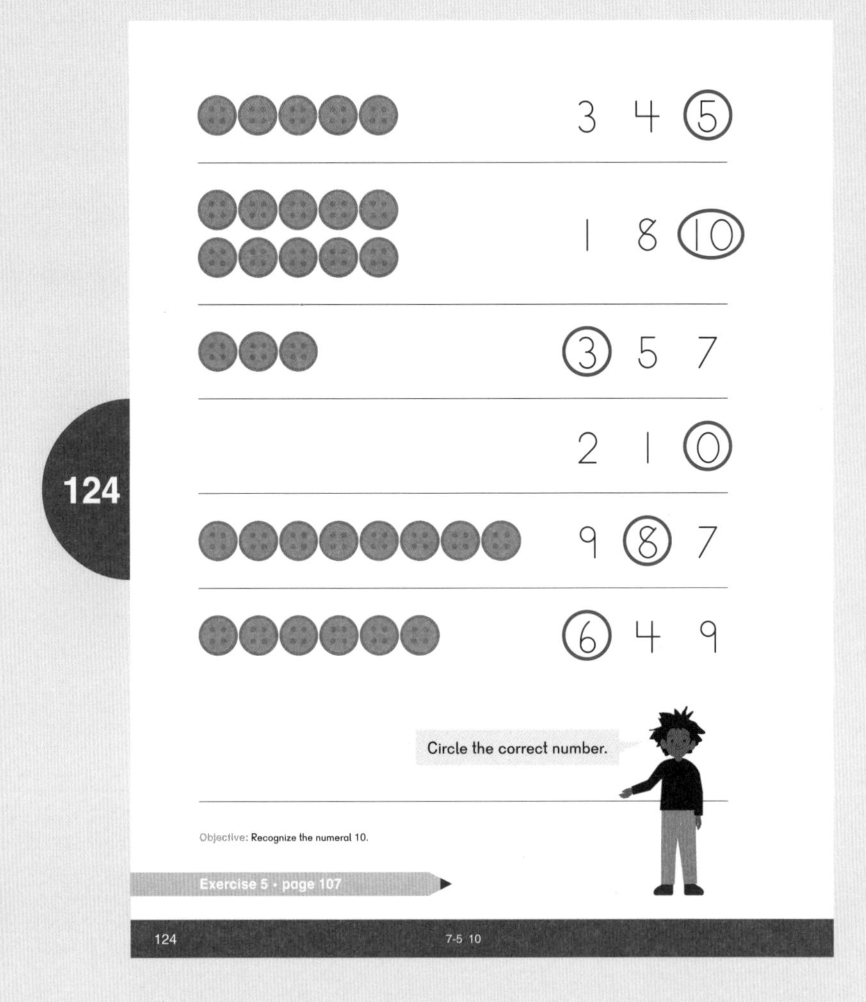

Small Group Center Play

- **Sort:** Offer buttons of different sizes, colors, and number of holes. Have students create several different patterns, each with 10 buttons.
- **Music:** Include drums. Encourage students to create different beats and have their friends copy the beats.
- **Dress-Up:** Include items of clothing for superheroes.
- **Kitten Yarn Art:** Tell students that kittens love to play with yarn. Provide yarn, glue, and pieces of paper with 10s written on them. Have students use yarn to cover the 10.

Exercise 5 • page 107

Extend Explore

What's Different?: Have students record their ideas about what makes the number 10 different from 0 to 9.

Materials: Recording device

Objective

- Review the numerals 0 to 10.

Lesson Materials

- Number Cards — Large (BLM) 0 to 10 – 1 set
- Number Cards (BLM) 0 to 10 – 1 set per student
- Optional snack: Finger sandwiches

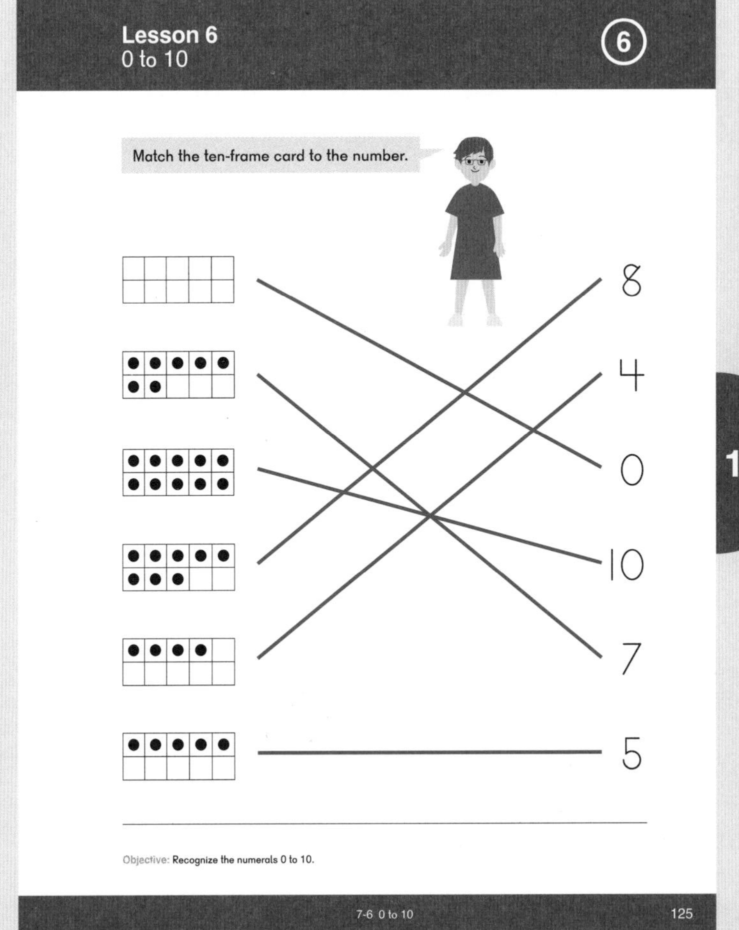

125

Explore

Take students for a walking tour of the inside or outside of the school. Have them point to any numeral they see and identify it.

Bring students back to class. Give each student a set of Number Cards (BLM), 1 to 10, not in numerical sequence. Sing the song from **Chapter 6: Lesson 2** with students (page 140 in this Teacher's Guide and page 93 in the textbook). Sing the song again, slowly, and each time a number word is sung have students hold up the correct card.

Hold up a Number Card — Large (BLM) and have students show you that many fingers, starting with their left pinkies, as quickly as they can. Tell them to show a fist when you hold up the "0." After several minutes, allow them to show the correct number of fingers a different way.

Learn

Have students look at page 125. Read Emma's direction and have them complete the task.

On page 126, have students tell you how many blue crayons there are, then draw a line from the crayons to the 3. Read Sofia's direction and have them complete the task.

Whole Group Play

That's My Number: Have students sit on chairs in a circle. Give each student a Number Card — Large, 0 to 10. Tell students that as you call a number, any students holding that number card must stand up, hold their card up in the air, shout, "That's my number!" and sit down. After all numbers 0 to 10 have been called once, have students exchange cards and repeat. This time, repeat some numbers. A sequence might be, 1, 3, 2, 9, 2, 2, 10, 4, 5, 8, 8, 6, 1, 7, 7, etc.

Materials: Fly swatters, Number Cards — Large (BLM) 0 to 10

Small Group Center Play

Patterns: Include crayons of different colors. Have students create several different patterns by drawing circles of different colors on paper.

Music: Include drums. Encourage students to create different beats and have their friends copy the beats.

Crayon Store: Include crayons and bags. Have students guess how many crayons will fit in a bag, then fill the bag to prove or disprove their theories.

Soap Crayons: Use grated white soap or powdered soap. Add water hot enough to melt the soap, then food coloring. Stir until blended. Have students mold the soap into numbers of their choice by putting a piece of wax paper over a number card and having them mold the soap in the shape of their chosen number.

Exercise 6 • page 109

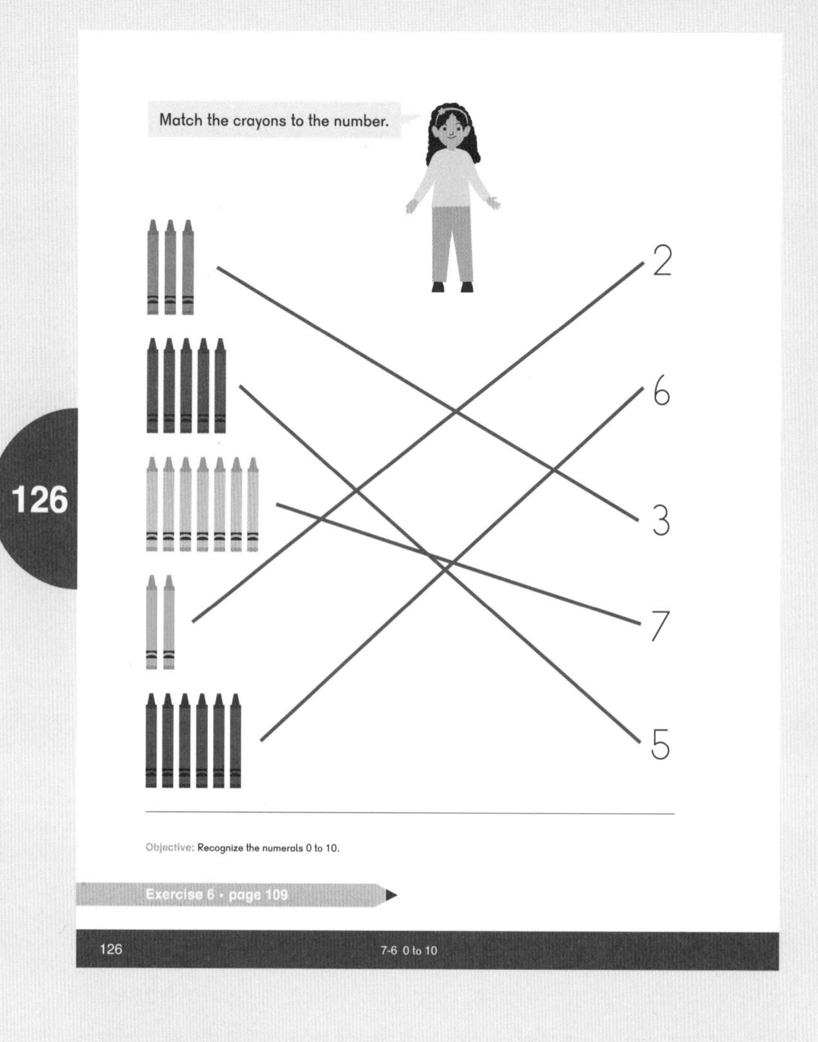

Extend Play

0 to 10 Memory: Use one set of Number Cards and one set of Ten-frame Cards (BLM) for each group of four students. Lay cards facedown to create a 4 × 5 grid. Player 1 starts by turning over two cards, trying to make a match of a ten-frame card and a number card showing the same number. If Player 1 finds a match, she puts those cards aside and tries again. If Player 1 does not find a match, she turns the cards back over and the next player gets a turn. The winner is the student who has the most matches when all cards have been matched.

Materials: Number Cards (BLM) 0 to 10, Ten-frame Cards (BLM) 0 to 10

Objective

- Match a given numeral 0 to 10 to a set containing that number of objects and to a ten-frame card representing the same.

Lesson Materials

- Painter's tape
- Number Cards (BLM) 1 to 10, 1 set per student
- Bags containing 1 to 10 small counters, 1 bag per student (for convenience, you may choose to make bags containing 1 have the same type/color of container, different type for 2, etc.)
- Optional snack: Bananas

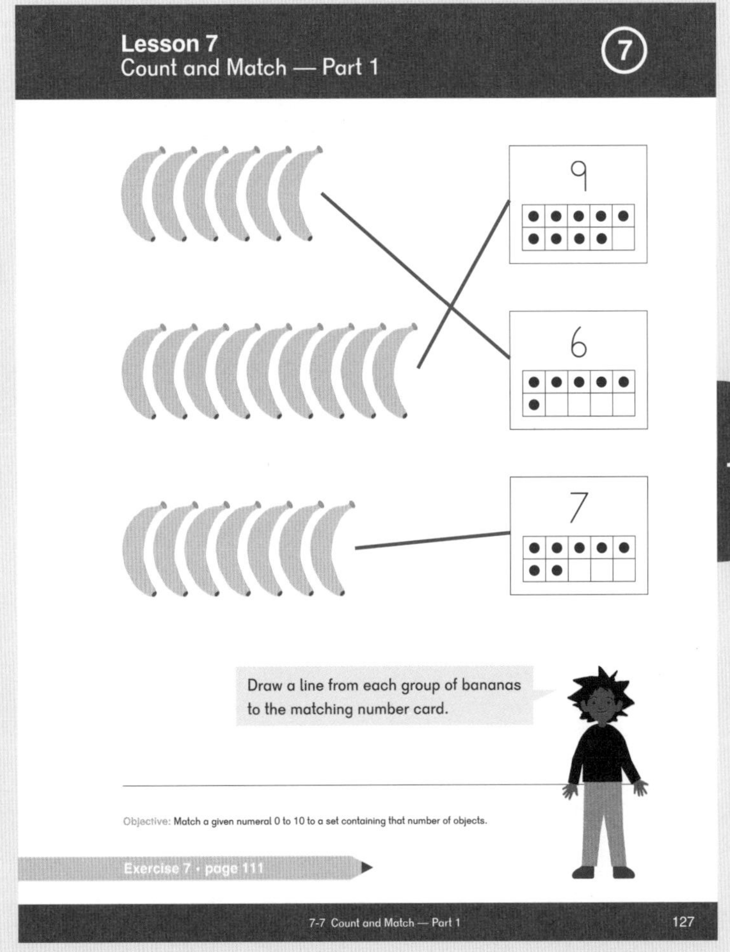

Prior to starting this lesson, create a large number path on the floor. The numbers should each be about the size of poster board.

Explore

Give each student a bag of counters. Have them count, place the counters back in the bag, then take their bag and sit near the matching number on the floor. Differentiate the number of counters based on students' understanding. If more than one student is on a number, students should count each other's counters to check. After all students are sitting, give each student a set of Number Cards (BLM) 1 to 10 and have them find the number that matches their counters.

Learn

Have students look at page 127. Read Dion's direction and have them complete the task.

Ask students questions, such as:

- Which was easier to count without touching, the bananas or the ten-frames?
- Why do you think that?
- How did you count the bananas?
- How did you count the ten-frames?
- Could you tell the number of any group(s) without counting? How did you know?
- Did you count the bananas differently from the ten-frames? Why or why not?
- Show me how you counted.

This kind of chat will help those who aren't seeing the organization of the frames as a way to help in counting.

Whole Group Play

What Can You Build?: Give each student a Number Card (BLM) with a number between 6 and 10. Have each student count out that number of blocks, then move to an area on the floor and begin to build something. Several students may decide to build together. Give them a few minutes, then have all students stop building.

Have students look at all of the constructions and compare them by saying such things as, "This is the tallest," "This is the shortest," "This one looks like it has more blocks than that one," etc. Then have students deconstruct their creations and build something different using the same materials.

Materials: Number Cards (BLM) 6 to 10 – 1 card per student, blocks

Small Group Center Play

Patterns: Include building blocks of different sizes and shapes. Have students create patterns of their choice using 10 pattern blocks for each.

Music: Include Number Cards (BLM) 1 to 10, facedown, and musical instruments. Have students draw a number and make that many beats on the drum or play that many notes.

Monkey Talk: Tell students that monkeys eat bananas. Have them pretend to be monkeys.

Banana Art: Include yellow paint and number cards, facedown. Have students draw a number card and paint that many bananas on art paper, then glue the numeral to their creations.

Match: Include Number Cards (BLM) 0 to 10 and Ten-frame Cards (BLM) 0 to 10. Lay out the cards in a grid and have students match a number card to a ten-frame card.

Exercise 7 • page 111

Extend Play

How Tall Can You Build?: Tell students that you want them to build the tallest structure possible using 10 wooden blocks of different sizes. Give them several minutes to create their structures, then talk about whose structure is the tallest. Have the builder of that structure discuss his or her strategy for building the structure.

Materials: Wooden blocks

Time for a Brain Break!

Your students have been learning a lot! Take a break and focus more on visualization and subitizing.

Visualization Activities

Copy My Card: Flash a dot card at the students and have them copy your card using counters on a paper plate. Repeat with other arrangements of three stickers, or with increasing numbers of stickers, up to 10, as students become proficient.

Materials: Index cards showing different arrangements of 3 or more dot stickers, 1 paper plate per student, two-color counters

Near or Far: Take your students outside or into a hallway where there is some distance between them and an object they can see. Stand beside the object and compare to their height, then walk away from the object and compare again. Ask them what they notice about the relative sizes of things they see that are close to them and things that are far away.

Goldilocks Story Time: After determining that things far away appear smaller, have students look at page 15 from **Dimensions Math® Pre-K Textbook A.** Have them tell the story "Goldilocks and the Three Bears" in their own words. Then have them imagine that Goldilocks turns around in the picture so that she is looking at the three bears and their house. Ask them why, in the picture, Papa Bear looks like he is almost the same size as the house.

Materials: Dimensions Math® Pre-K Textbook A

How Many Windows?: Using the same picture of Goldilocks, have the students say how many windows are on the front of the house. Then see if they can say how many windows are on the side of the house quickly without counting.

Lesson 7
Size — Part 1

7

15

Objective: Use a familiar story to recognize the color orange and learn the words "big," "little," "small," and "large."

1-7 Size — Part 1 15

Objective

- Practice matching a given numeral 0 to 10 with a drawn picture containing that number of objects.

Lesson Materials

- Number Cards (BLM) 1 to 10, 1 set per student
- 2 sets of pictures from magazines or newspapers showing different numbers of people, animals, or objects
- Optional snack: Animal crackers

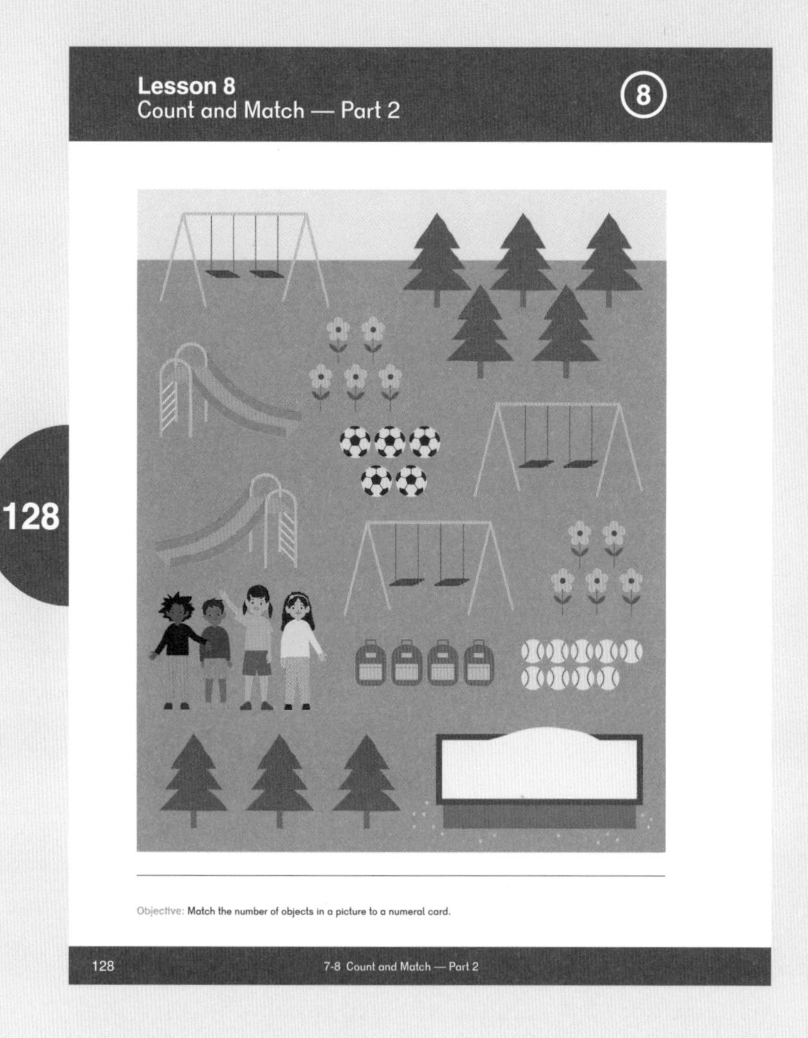

Explore

Give each student a set of Number Cards (BLM) 1 to 10.

Hold up pictures showing different numbers of people, animals, or objects. Students must find the correct number card (BLM) and hold it up.

Learn

Have students look at page 128 and discuss the picture. The activity for this page will be done in small groups. Ask students, "What questions could we ask about this picture?" Then have them count the sets of objects on the page and hold up the matching number cards.

Have students look at page 129. Read Emma's direction and have them complete the task.

Have students look at page 130. Read Mei's direction aloud and have them complete the task.

Whole Group Play

Roll It: Create paper plate number necklaces for each student by writing a number 1 to 10 on a paper plate and using yarn to go around the neck. Have students sit in a circle and roll a ball to each other. Before rolling the ball, the roller must call a number to alert the student wearing that number. If there is more than one student wearing that number, the roller can roll to either student. If possible, take them outside and have them play catch.

Materials: Paper plates, yarn, ball

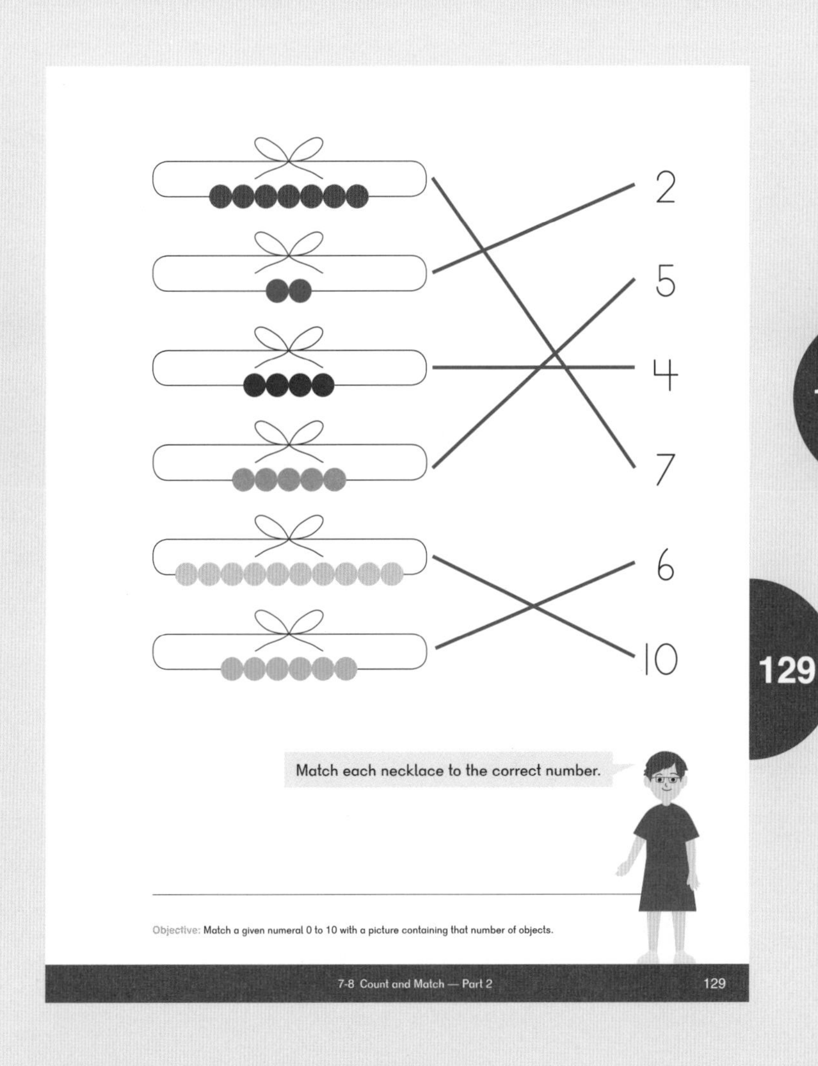

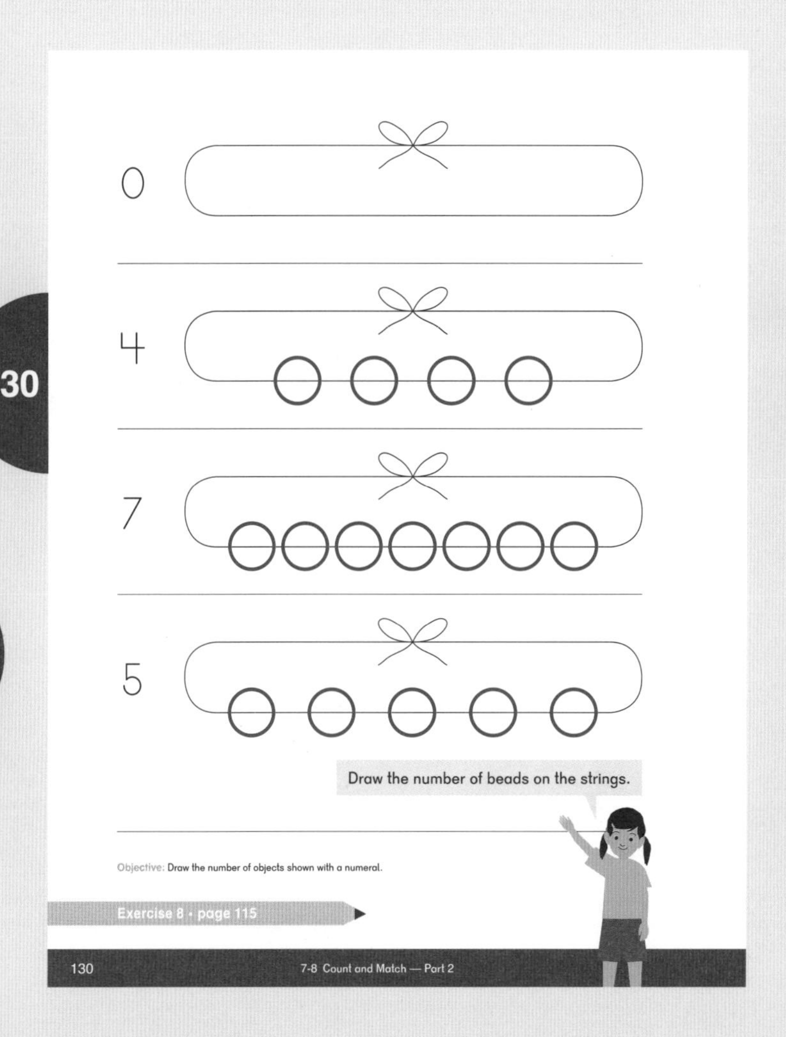

Small Group Center Play

Patterns: Include beads and string. Have students create bead patterns of their choice using up to 10 beads.

Music: Include Number Cards (BLM) 1 to 10 facedown and musical instruments. Have students draw a number and make that many beats on the drum or play that many notes.

Kitchen: Include different numbers of play foods and number cards 1 to 10. Have students count the food items of each type and match a number card to it.

Sand Art: Provide small boxes of sand, medicine droppers filled with food coloring, and small transparent bottles. Have students pick a number card to determine how many colors to use for their creations.

Match It: For each group of three or four students, set out a 4 × 5 grid of cards using Number Cards (BLM) 1 to 10 and Picture Cards (BLM) 1 to 10. Cards should be faceup. Students take turns matching a picture card to a number card.

Exercise 8 • page 115

Extend Play

Number Collage: Before class, prepare art paper by writing a number 1 to 10 on it. During class, give each student a piece of the prepared art paper. Tell them that their task is to find magazine pictures that will match the numerals on the art paper, cut out the pictures, and glue them to the paper near the matching numeral.

Materials: Art paper, magazines, scissors, glue

Objective

- Practice concepts introduced in this chapter.

Lesson Materials

- Optional snack: Flowers with baby carrots or bell pepper slices for petals and cucumber slices for center

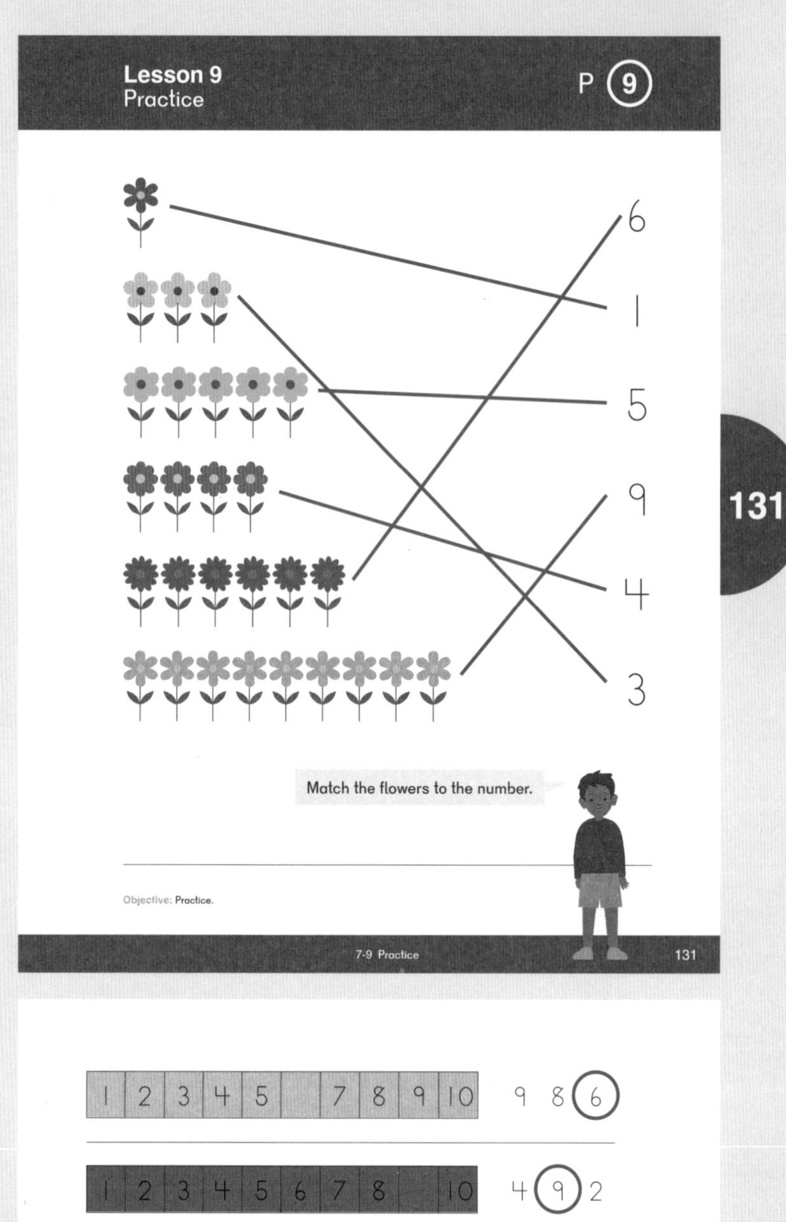

For the **Practice,** read the directions and speech bubbles on each page and have students complete the task.

Whole Group Play

Number Hunt Walk: Take students for a walk around the inside or outside of the school building. Consider making "binoculars" out of two toilet paper rolls taped together to go looking for numbers. Have students point to any numerals they see and identify them.

Materials: Empty toilet paper rolls, tape

Small Group Center Play

Patterns: Include different objects for patterning. Have students create patterns of their choice.

Music: Allow students to play with instruments freely.

Dress-Up: Give students a Number Card (BLM) 6 to 10 and have them use that many items of clothing and accessories.

Flower Shop: Include artificial or paper flowers and tall, thin containers. Have students create floral arrangements. Write a numeral 1 to 10 on each vase. Have students match their arrangement to a numeral by placing the flowers in the correct container.

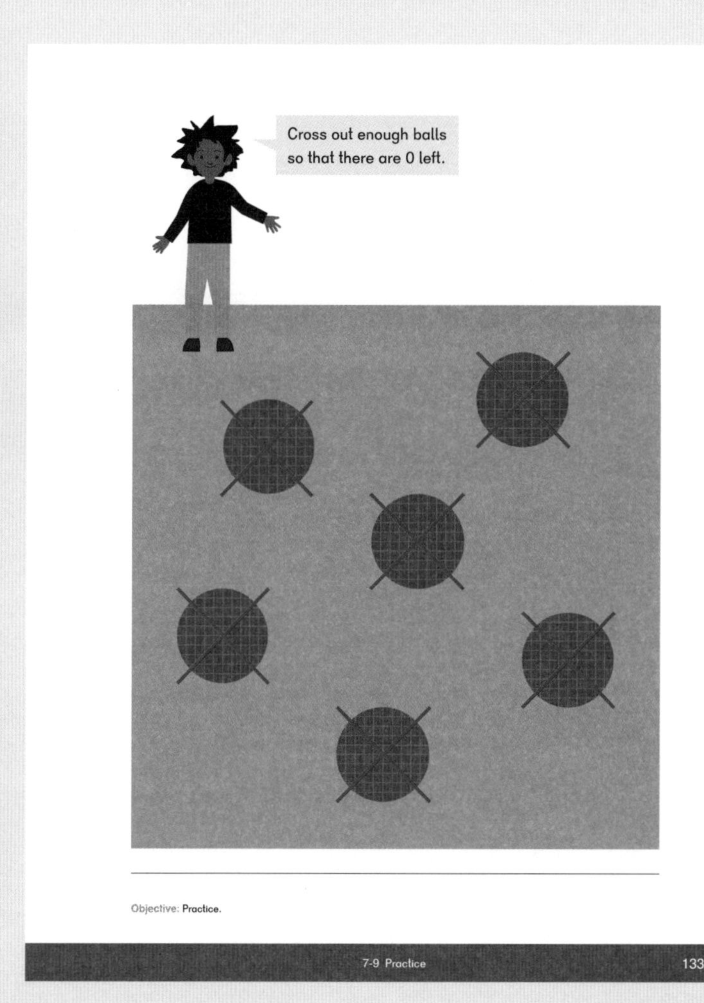

133

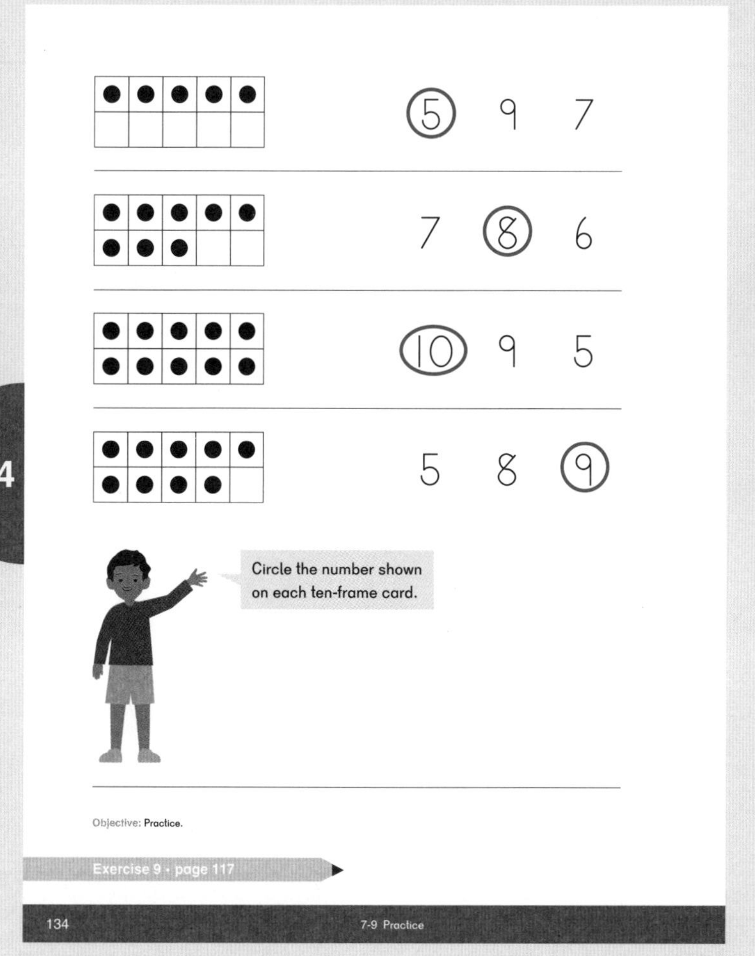

134

Exercise 9 • page 117

Extend Play

Computer Time: Have students type numerals and identify them on the screen after typing them.

Materials: Computers

Notes

Exercise 1 • pages 99 – 100

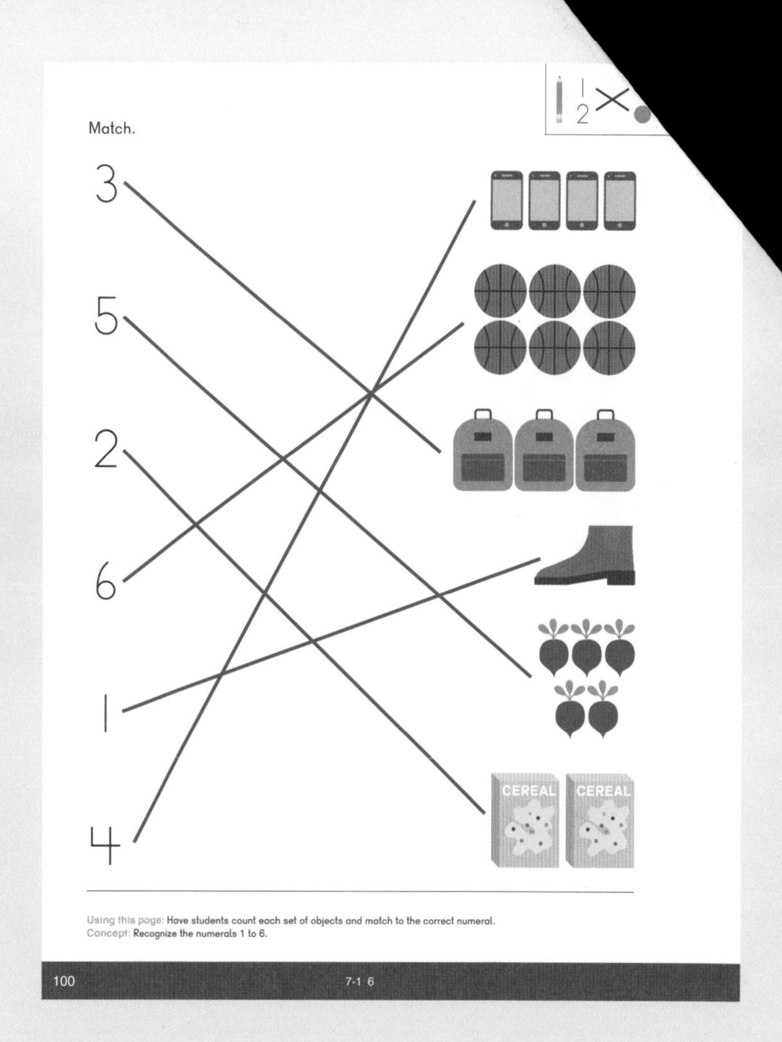

Exercise 2 • pages 101 – 102

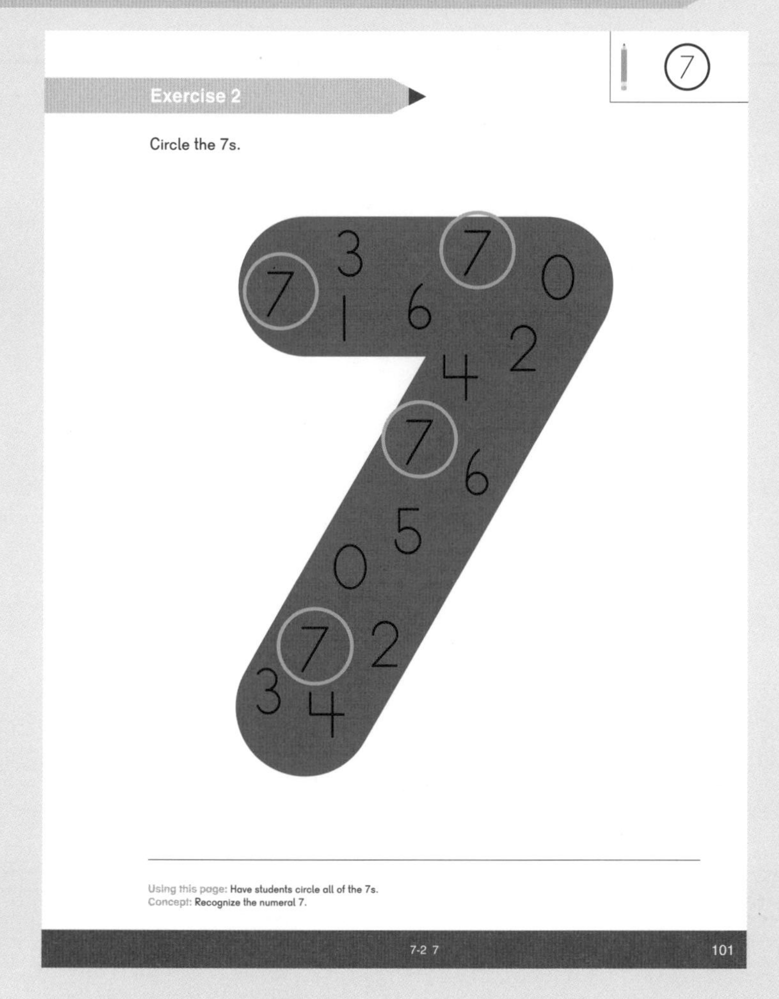

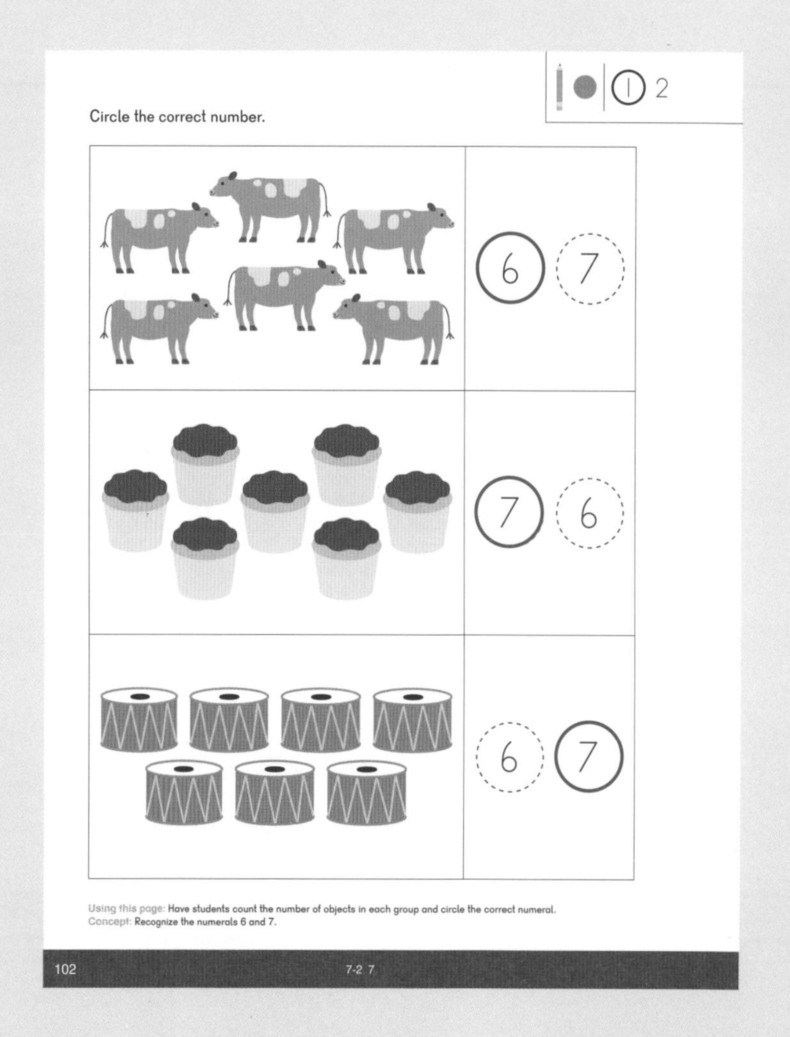

...ges 103 – 104

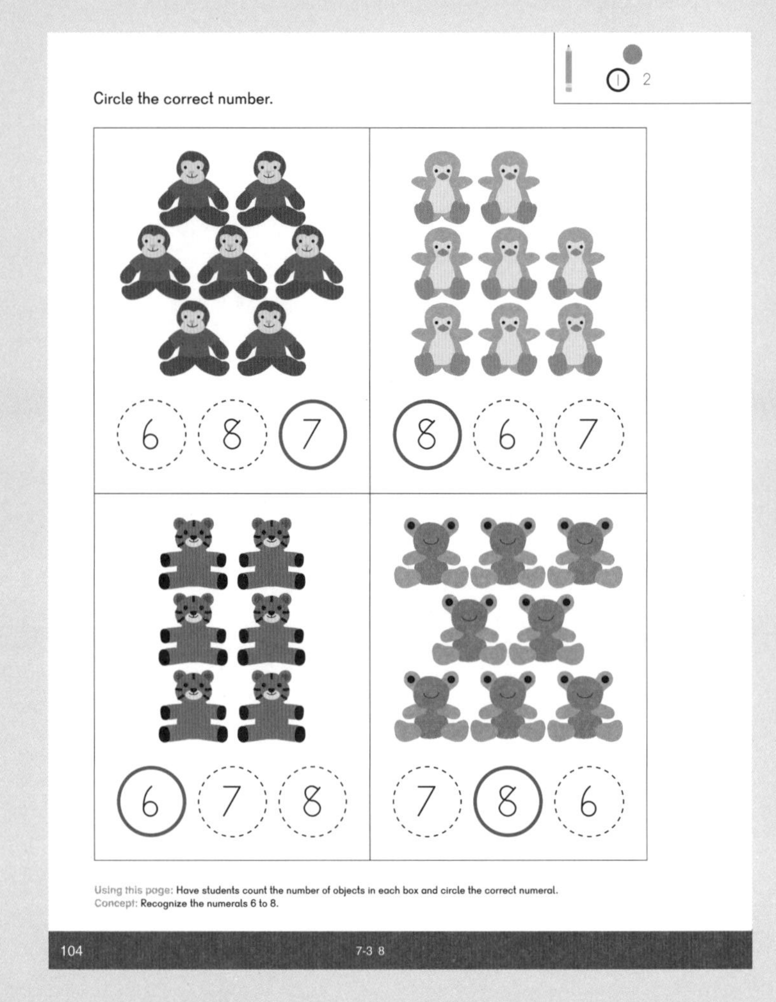

Exercise 4 • pages 105 – 106

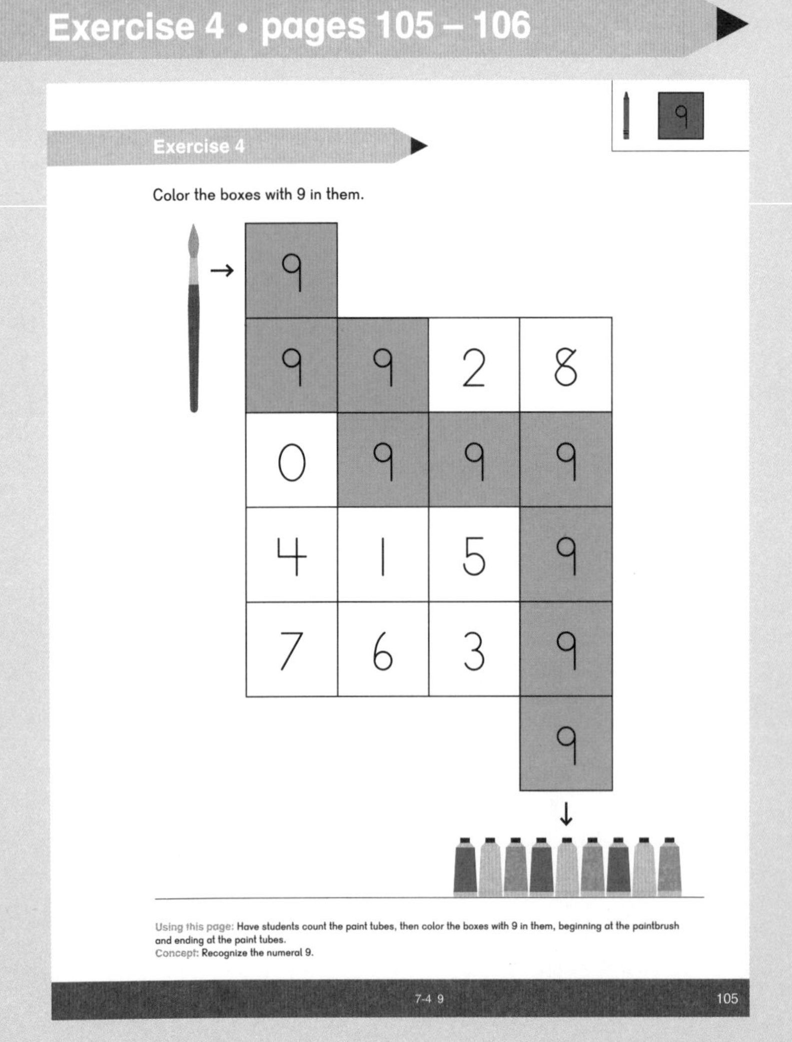

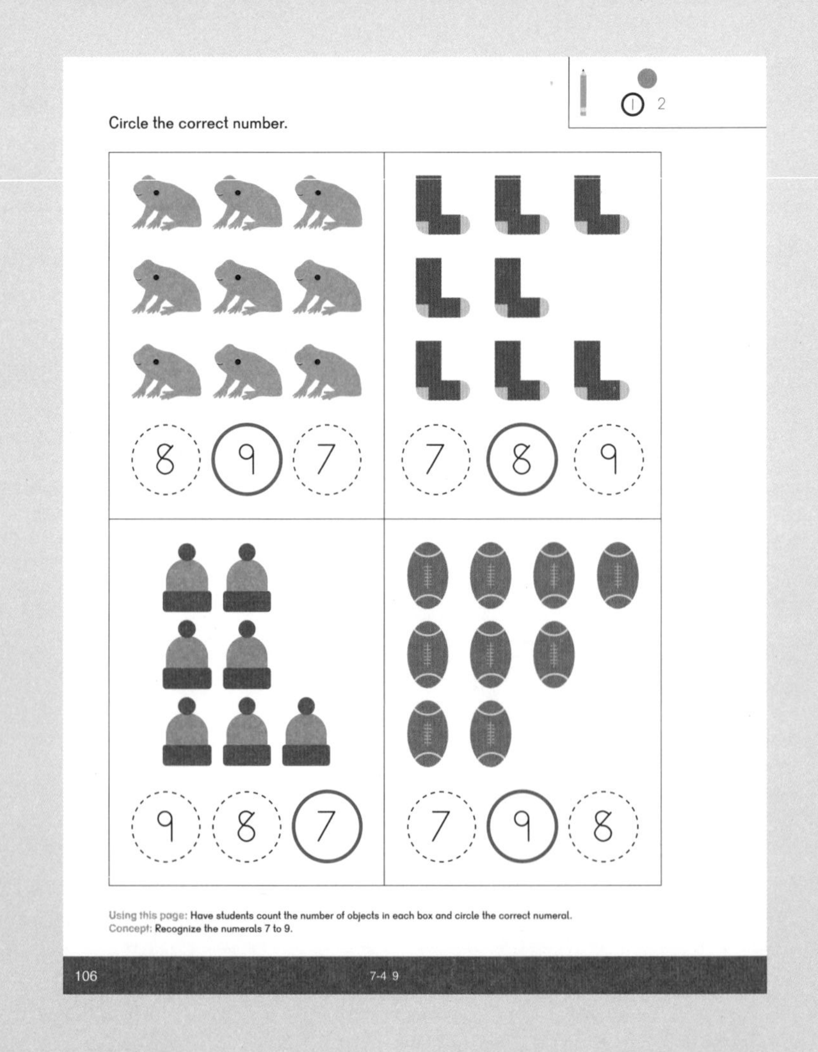

Exercise 5 • pages 107 – 108

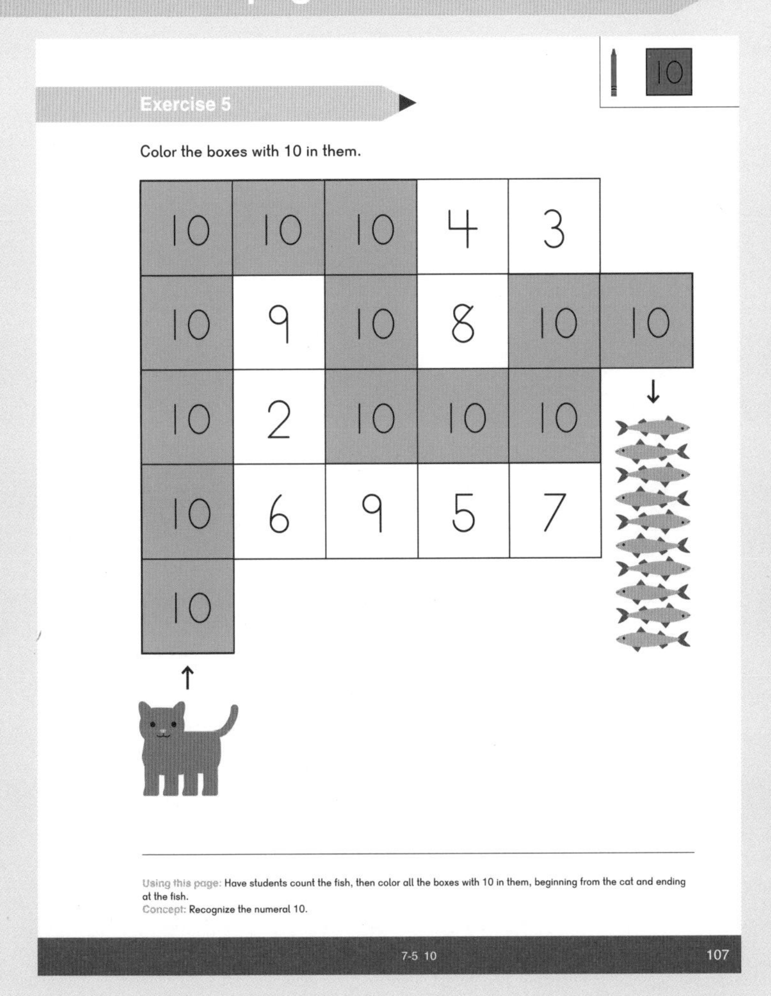

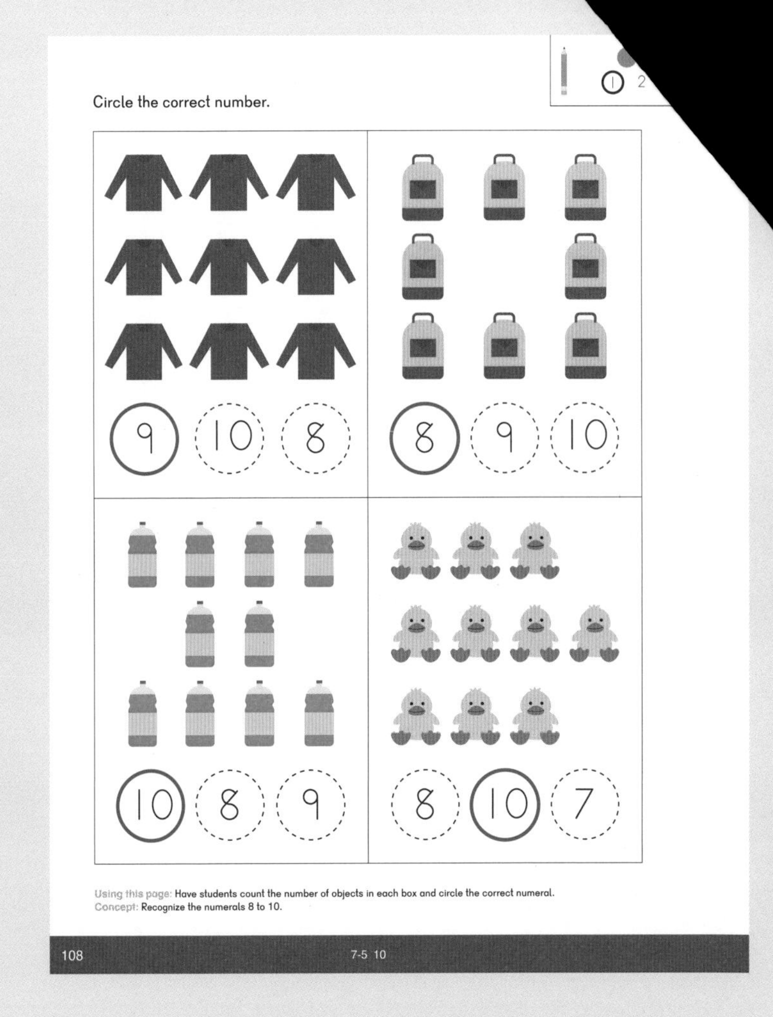

Exercise 6 • pages 109 – 110

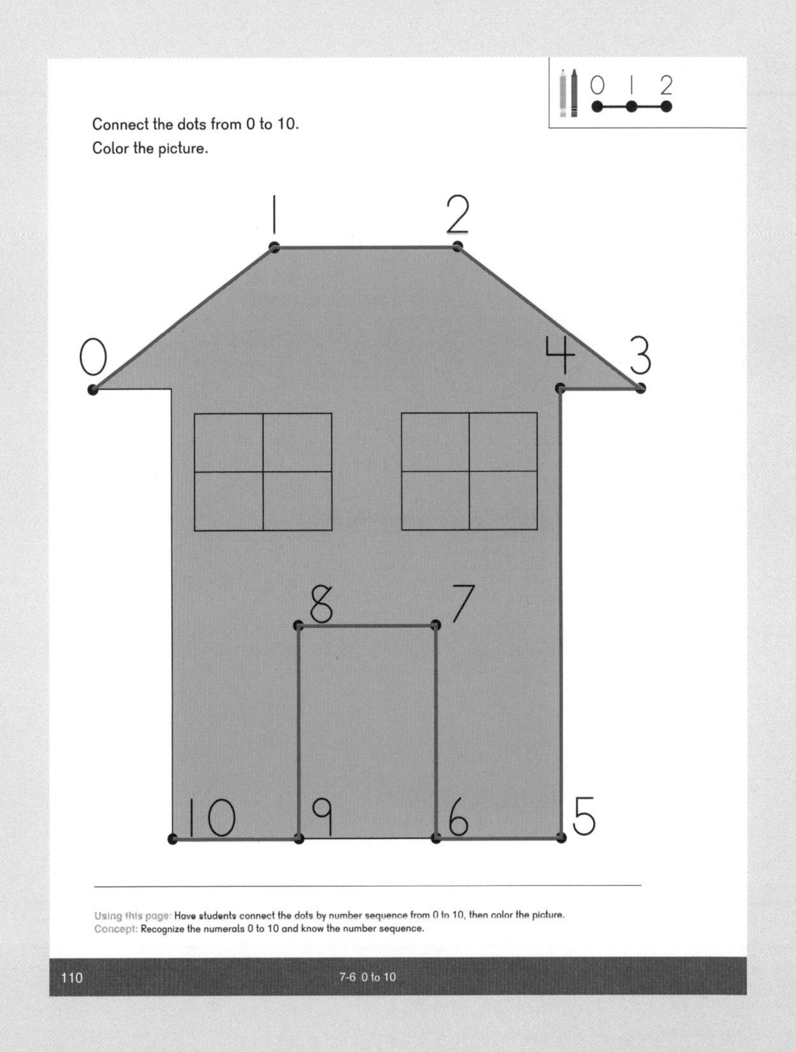

...ges 111 – 114

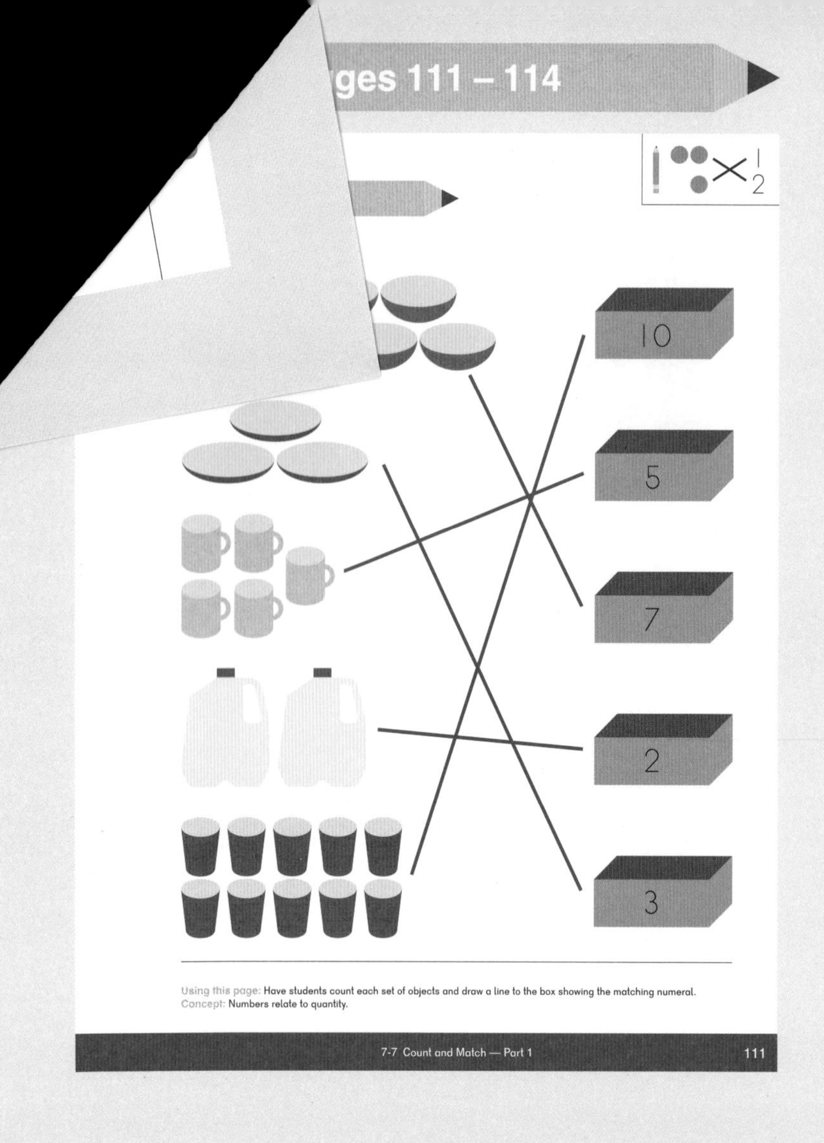
10
5
7
2
3
Using this page: Have students count each set of objects and draw a line to the box showing the matching numeral.
Concept: Numbers relate to quantity.
7-7 Count and Match — Part 1
111

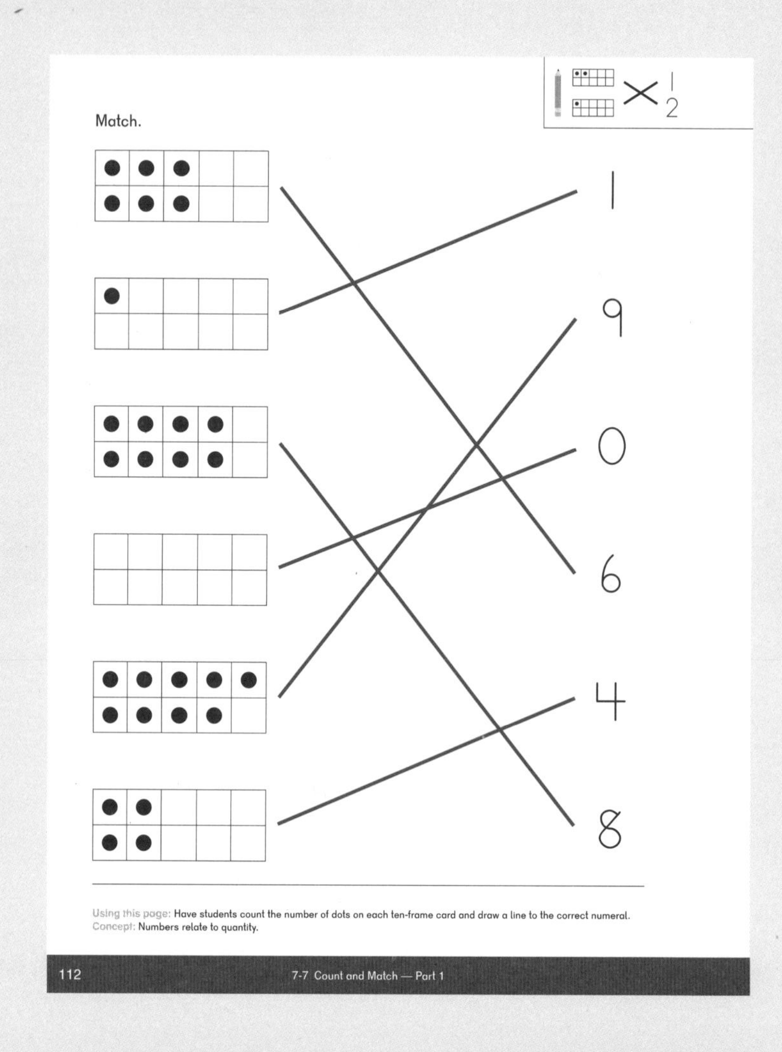
Match.
1
9
0
6
4
8
Using this page: Have students count the number of dots on each ten-frame card and draw a line to the correct numeral.
Concept: Numbers relate to quantity.
112
7-7 Count and Match — Part 1

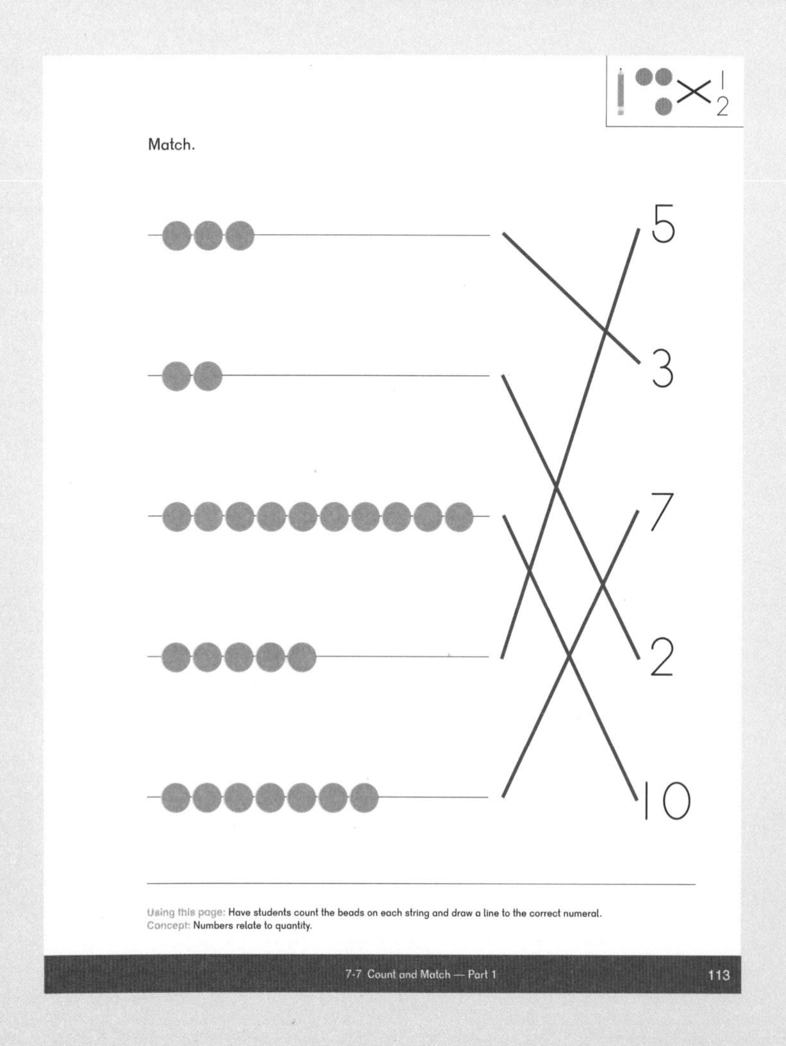
Match.
5
3
7
2
10
Using this page: Have students count the beads on each string and draw a line to the correct numeral.
Concept: Numbers relate to quantity.
7-7 Count and Match — Part 1
113

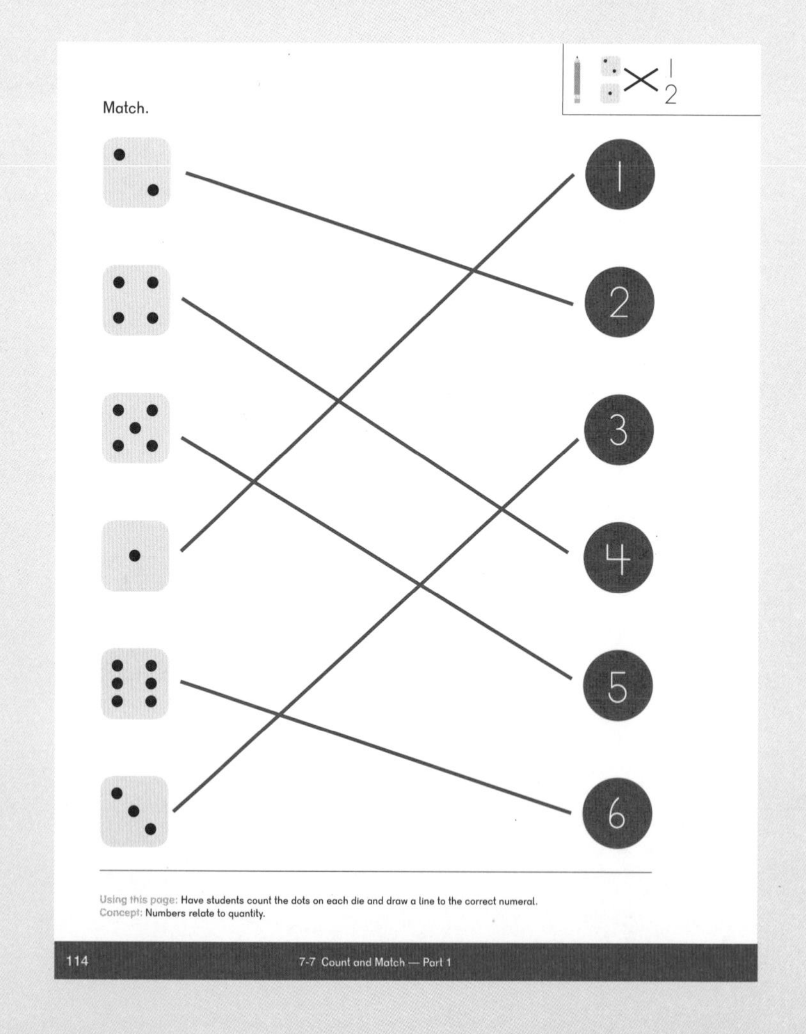
Match.
1
2
3
4
5
6
Using this page: Have students count the dots on each die and draw a line to the correct numeral.
Concept: Numbers relate to quantity.
114
7-7 Count and Match — Part 1

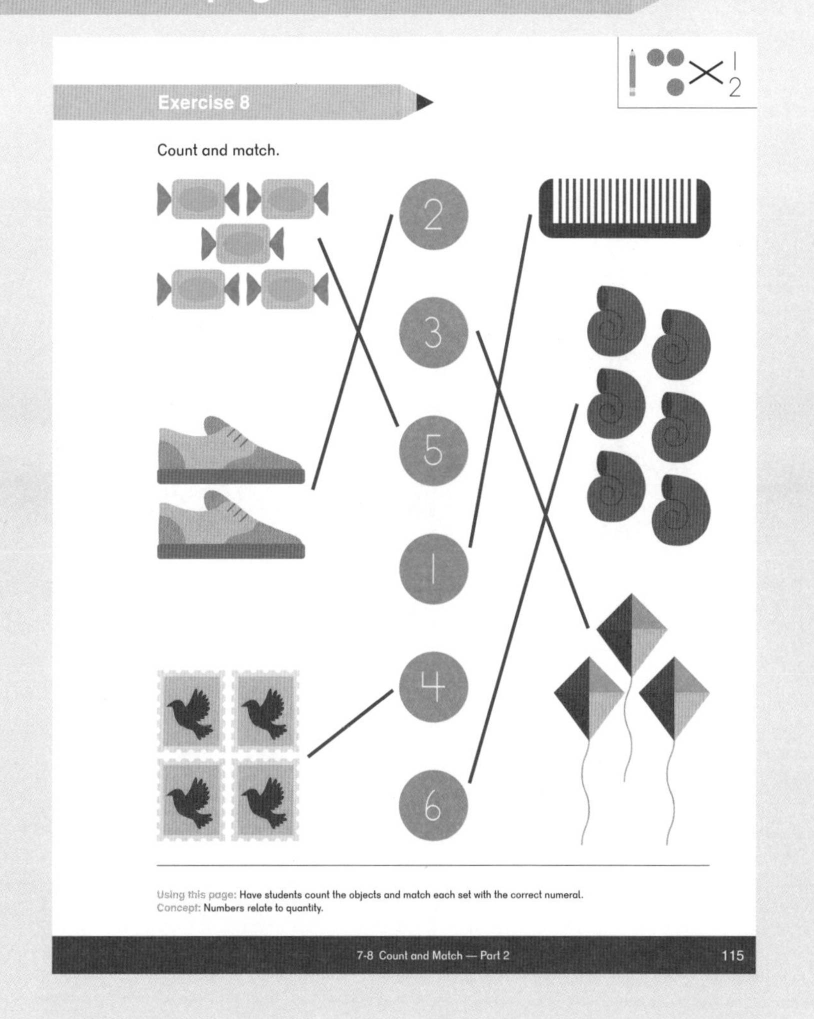
Exercise 8
Count and match.
2
3
5
1
4
6
Using this page: Have students count the objects and match each set with the correct numeral.
Concept: Numbers relate to quantity.
7-8 Count and Match — Part 2
115

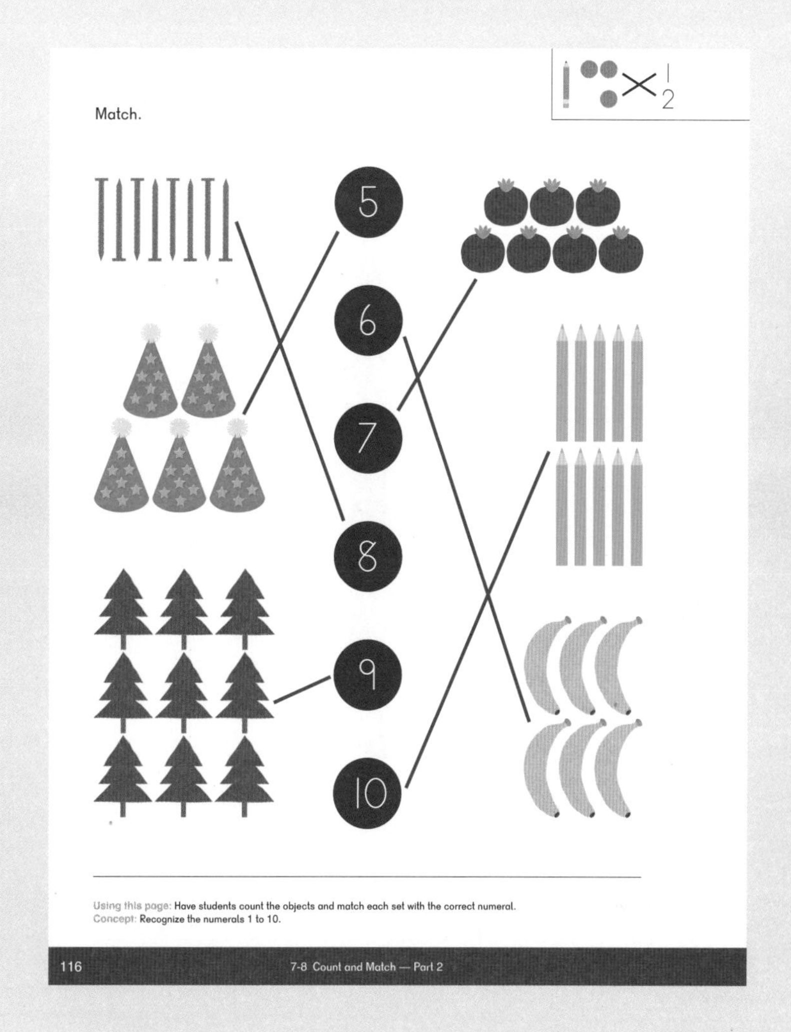
Match.
5
6
7
8
9
10
Using this page: Have students count the objects and match each set with the correct numeral.
Concept: Recognize the numerals 1 to 10.
116
7-8 Count and Match — Part 2

Exercise 9 • pages 117 – 120

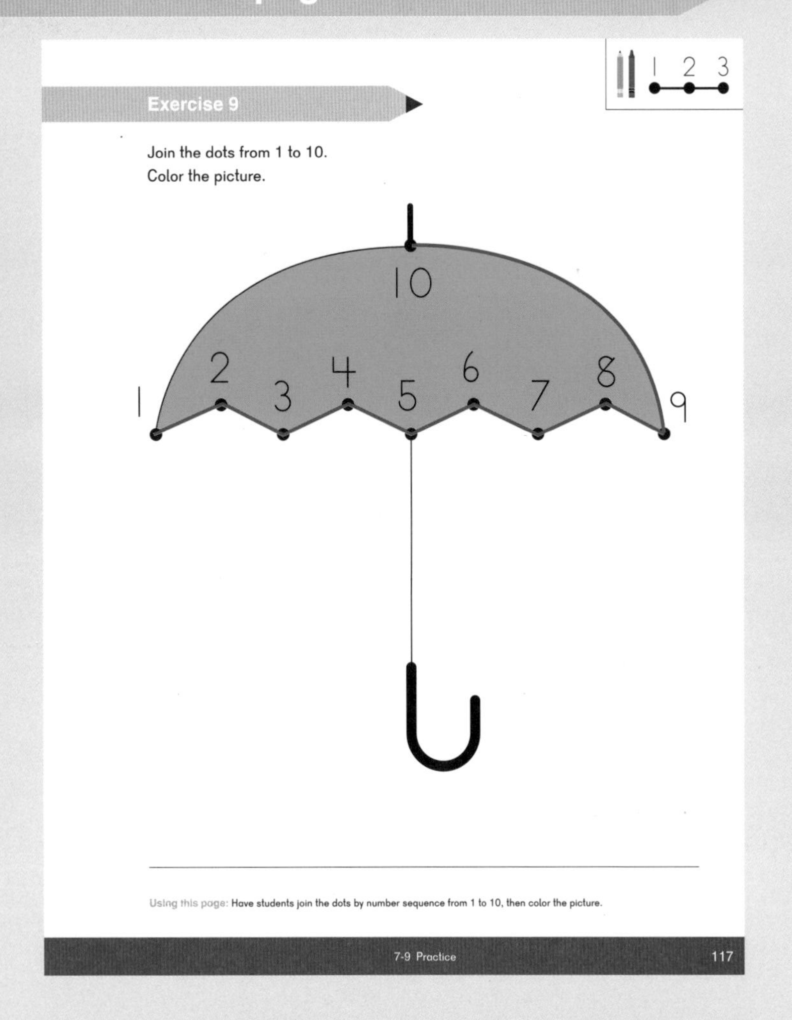

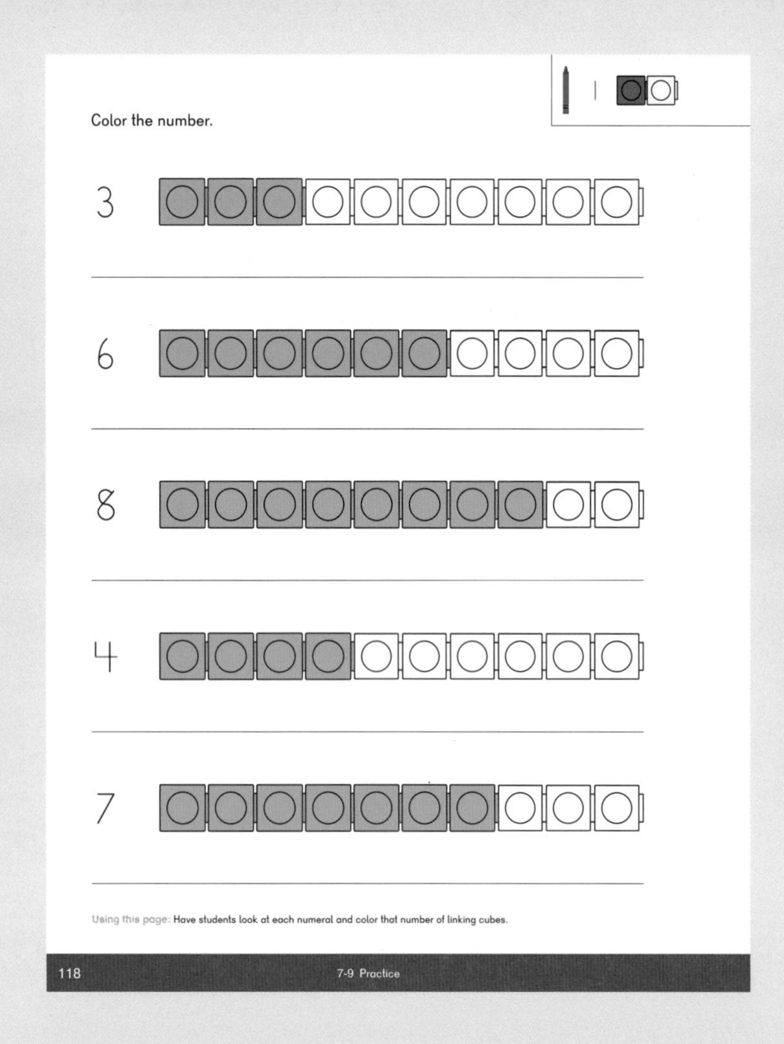

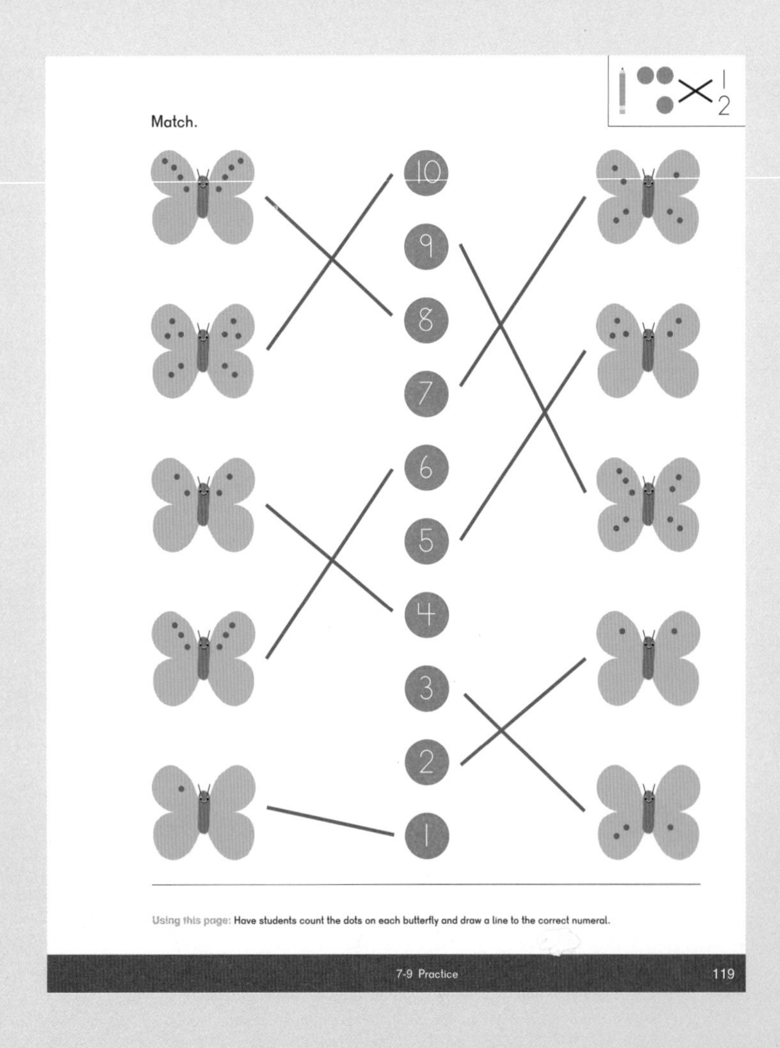

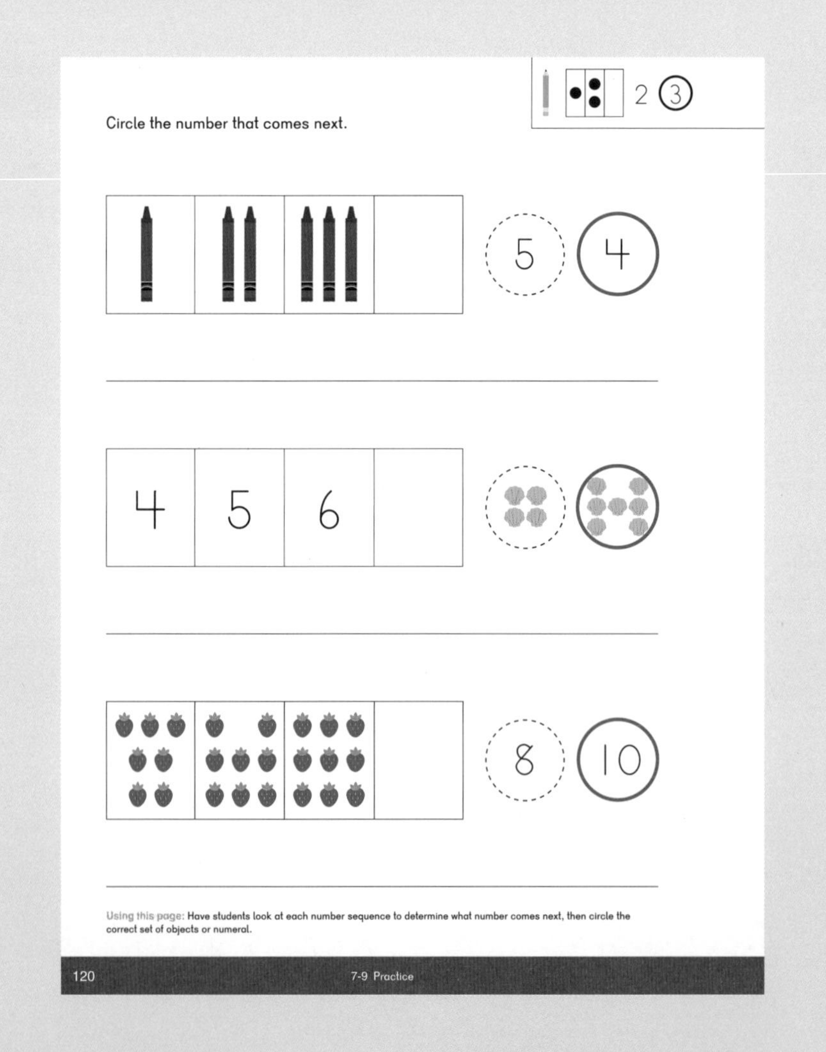

All Blackline Masters used in the guide can be downloaded from dimensionsmath.com.
This lists BLM used in the **Explore** and **Learn** sections.
BLMs used in **Activities** are listed in the Activity Materials within each chapter.

Blank Five-frame	**Chapter 4:** Lesson 9, Lesson 10, Lesson 11, Lesson 13 **Chapter 5:** Lesson 4, Lesson 5, Lesson 6
Blank Ten-frame	**Chapter 6:** Lesson 4, Lesson 6, Lesson 7, Lesson 10 **Chapter 7:** Lesson 2, Lesson 3, Lesson 6
Counting Up to 5 Template	**Chapter 4:** Chapter Opener, Lesson 9
Cuisenaire Rods	**Chapter 6:** Lesson 12
Dot Cards	**Chapter 4:** Lesson 12 **Chapter 5:** Lesson 5, Lesson 6, Lesson 7 **Chapter 7:** Lesson 1
Five-frame Cards	**Chapter 4:** Lesson 9 **Chapter 5:** Lesson 4, Lesson 6
Four-column Template	**Chapter 4:** Lesson 4
Individual Number Paths	**Chapter 5:** Lesson 3
Keyboard Template	**Chapter 4:** Lesson 10 **Chapter 6:** Lesson 5, Lesson 8
Number Cards — Large	**Chapter 5:** Lesson 1, Lesson 2 **Chapter 7:** Lesson 1, Lesson 2, Lesson 3, Lesson 4, Lesson 5, Lesson 6, Lesson 7, Lesson 8
Number Cards — Small	**Chapter 4:** Lesson 10 **Chapter 5:** Lesson 1, Lesson 4, Lesson 5, Lesson 6, Lesson 7 **Chapter 7:** Lesson 1, Lesson 5, Lesson 6, Lesson 7, Lesson 8, Lesson 9
Number Cards FFR — Large (Five-frame Representation)	**Chapter 5:** Lesson 2
Number Cards FFR — Small (Five-frame Representation)	**Chapter 7:** Lesson 1, Lesson 3, Lesson 4
Octopus Template	**Chapter 6:** Lesson 8

Picture Cards	**Chapter 5:** Lesson 5, Lesson 6 **Chapter 7:** Lesson 8
Table Setting	**Chapter 1:** Lesson 9
Rhyming Rebus	**Chapter 1:** Lesson 1
Ten-frames 0 to 10	**Chapter 6:** Chapter Opener, Lesson 8, Lesson 12 **Chapter 7:** Chapter Opener, Lesson 2, Lesson 4, Lesson 6, Lesson 7, Lesson 8
Vegetable Cards	**Chapter 6:** Lesson 6
Who Needs Teeth?	**Chapter 1:** Lesson 4